Cracking the Aging Code

ALSO BY JOSH MITTELDORF

Aging Is a Group-Selected Adaptation (a companion volume to *Cracking the Aging Code* for academic biologists)

ALSO BY DORION SAGAN

Cosmic Apprentice: Dispatches from the Edges of Science

Biospheres: Metamorphosis of Planet Earth

Into the Cool: Energy Flow, Thermodynamics, and Life (with Eric D. Schneider)

Up from Dragons: The Evolution of Human Intelligence (with John Skoyles)

Death/Sex (with Tyler Volk)

Origins of Sex: Three Billion Years of Genetic Recombination (with Lynn Margulis)

Microcosmos: Four Billion Years of Microbial Evolution (with Lynn Margulis)

Mystery Dance: On the Evolution of Human Sexuality (with Lynn Margulis)

What Is Life? (with Lynn Margulis)

What Is Sex? (with Lynn Margulis)

Garden of Microbial Delights: A Practical Guide to the Subvisible World (with Lynn Margulis)

Dazzle Gradually: Reflections on the Nature of Nature (with Lynn Margulis)

Acquiring Genomes: A Theory of the Origin of Species (with Lynn Margulis)

Lynn Margulis: The Life and Legacy of a Scientific Rebel (editor)

Cracking the Aging Code

*The New Science of Growing Old—
And What It Means for Staying Young*

JOSH MITTELDORF
and
DORION SAGAN

 Printed in the United States of America. For information, address Flatiron Books, 175 Fifth Avenue, New York, N.Y. 10010.

www.flatironbooks.com

The Library of Congress Cataloging-in-Publication Data is available upon request.

ISBN 978-1-250-06170-6 (hardcover)
ISBN 978-1-250-06172-0 (e-book)

First Edition: June 2016

10 9 8 7 6 5 4 3 2 1

"Go over the heads of the scientists," she said,

"take your case directly to the people.

That's what Darwin did!"

To my mother,
Harriet Mitteldorf
(1922–)

JJM

Contents

'Tis not that Dying hurts us so—
'Tis Living—hurts us more—
But Dying—is a different way—
A Kind Behind the Door—

—EMILY DICKINSON

Now flip, flop, and fly, I don't care if I die.

—BIG JOE TURNER,
A KANSAS CITY BLUES CROONER
AND FOUNDER OF ROCK AND ROLL

Over my dead body.

—GEORGE S. KAUFMAN,
SUGGESTING HIS EPITAPH

PREFACE

What This Book Is About

What Is Aging?

Scientists sometimes use the word ***senescence.*** *The common meaning is a deterioration in many body functions that comes with age. Sometimes we will use the demographer's definition: an increased risk of death with passing time.*

It is a common belief that aging is inevitable and universal. Not on your life!

During the twentieth century, medical technology took enormous strides toward the conquest of infectious disease and recovery from trauma. With hygiene and sanitation, then antibiotics in the 1930s and an ever-expanding arsenal of vaccines, many plagues of the past have been vanquished: polio, syphilis, whooping cough, diphtheria, and cholera were once feared as a death sentence, and now they are footnotes in the mortality statistics. The diseases that remain are all associated with aging; diabetes, arthritis, and osteoporosis are growing, and the Big Three killers are cardiovascular disease, cancer, and Alzheimer's. Tens of billions of dollars have been spent on medical research over several decades trying to conquer these diseases with the same approach that succeeded so well for infectious disease.

That approach has been to work with the body, to stimulate the body's growth and inborn strength, to buttress its natural defenses. Even the reductionist tradition of Western, allopathic medicine has been influenced by the philosophy of natural medicine, working with the body instead of attempting to overpower it with technology or drugs. But what the doctors do not yet realize is that they are working with a suicidal patient.

Suicide Genes

"Man overboard!"

You run to the railing and throw him a lifeline. If only you can get the buoy within his reach, you might pull him back to safety . . .

Good toss! The life preserver is right in front of him, but he is not taking it. Is he just too weak? Has he lost the will to live? You call to him. "Go away," he answers. "Leave me alone!" And now your understanding of his situation is changed.

He didn't fall off the boat—he flung himself into the sea. To save a man from drowning is one thing; to dissuade him from suicide will require a different approach. In the immediacy of the crisis, you might take advantage of his exhaustion, jump in the water, overpower him, and carry him forcibly to safety. But next week, he might make another attempt. To help this man, you will have to get to know him, learn what is important to him, understand why he wants to kill himself, and convince him to choose life.

Doctors today are trying to help a body that does not want to be helped. Efforts to restore the body's natural balance won't work, because as we age, the body's natural metabolism is bent on self-destruction. Attempts to bolster the body's natural defenses are doomed to failure because the natural defenses are slowly being shut down with age.

Progress can be made against the Big Three diseases, and aging itself can be abated, but a different approach is required. We must be willing not just to assist the body but to coax, cajole, and even fight with it when appropriate. We must learn more about hormones and the signaling language that regulates metabolism. We must whisper the word "youth" in the body's own native language of biochemistry, a language as yet still somewhat foreign to us. But this is the language in which the entire life plan is spelled out, from development in the womb to aging and death.

Selfish Genes

The central idea of this book is that aging is built into our bodies. Aging doesn't just happen but is regulated and controlled by our genes. Our self-destruction is scheduled as much as is our development in childhood or our sexual development at puberty. Growth, puberty, and aging all unfold on a schedule programmed into the regulatory segments of our DNA.

But to an evolutionist, these things are not the same at all. A strong body helps us to survive and preserve ourselves. Sexual development is necessary for reproduction. These things are good for us, good for the prospects of our genes. They help our genes to be passed along, to prevail, and to spread in succeeding generations. This fits well with the idea of natural selection, the fundamental premise of Darwinian evolution. It is easy to understand how genes for growth and sexual development evolved. They are "selfish genes," because they help the body, their vehicle, and so they help themselves.

But aging is about weakening and dying. It cannot be good for the body. Genes may cause aging, but aging cannot promote the prospects of those genes; quite the contrary, aging ends the careers of the very genes that cause it. This does not make sense from the prevailing evolutionary perspective. Why would selfish genes kill their bodies? But in fact multiple lines of evidence suggest that is exactly what they are doing. As we get older, our genes turn against us, killing healthy nerve cells and muscle cells, permitting the thymus gland to wither away, which undermines our immune system. We consider this self-destruction normal, but in fact, it doesn't apply to all species. There are a few animals and many plants that do not age. And although selfish gene theory cannot explain why our bodies do this, another theory—the Demographic Theory described in chapter 8—does. The self-destruction we take for granted is in fact under evolutionary control. These evolutionary "suicidal tendencies," while they do not make us as individuals "live long and prosper," turn out to have a vital evolutionary function.

These suicide genes are the opposite of selfish genes. In the course of evolution, genes for aging must have fought an uphill battle against natural selection. How could aging have evolved?

This is a question that has been asked over and over since Darwin first skirted the issue 150 years ago. In fact, there was no mention at all of life span or aging in the first edition of *On the Origin of Species* (1859). Skeptics confronted him with the great range of life spans in nature. Why have life spans not evolved ever longer? Isn't that what we should expect from your theory? Darwin's answer in later editions was uncharacteristically vague and unconvincing. Although many volumes and thousands of articles have been written on the subject ever since, there are really only three kinds of answers.

Three Evolutionary Stories About Aging

Answer number one is that there really is no aging in nature. Animals in the wild do not live long enough to die of old age, because they die of other things first. It's a highly competitive world out there, "red in tooth and claw."* Organisms are ever at risk from predators, accidents, and starvation. We are aware of aging when we look at domesticated animals protected from their natural enemies, and of course civilization has afforded humans a safety that is completely uncharacteristic of our evolutionary history. Aging is an artifact of protected environments, and in the natural world where evolution runs its course, aging doesn't exist—so there is nothing to explain.

Answer number two is that natural selection encounters trade-offs and makes compromises. The body is doing its best to repair itself and forestall damage, but there are other priorities that keep it from doing a perfect job. In youth, the body allows itself to gradually accumulate damage, even though this eventually will prove fatal, because neglecting the infrastructure frees up resources for survival and reproduction.

Answer number three is that even though aging is bad for the individual, it is important for the community. Aging creates opportunities for

* From an 1849 poem by Alfred, Lord Tennyson, poet laureate of Great Britain and Ireland, in memory of his friend Arthur Henry Hallam.

the young and thus it promotes population turnover for adaptive change. Another communal benefit of aging is the stabilization of populations. Aging levels the death rate so individuals don't all die at once, as in famines and epidemics.

Answer number one (aging is an artifact of protected environments) has been thoroughly discredited by population studies of animals in the wild. It has been established that, indeed, animals in nature live long enough for aging to matter. Answer number two (trade-offs and compromises) depends crucially on the assumption that the body can only do a good job of one thing at a time. Although this seems to be true in particular cases, there is no reason to think that a strengthening in one area must always come at the expense of a weakening in another, and indeed there are abundant counterexamples. There is a lot of genetic and experimental evidence in favor of answer number three (communal benefit), but deeply entrenched evolutionary theory has created a strong prejudice against it.

This prejudice is a deep vein in our culture of individual competition. In economics, it is the myth of free markets. Socially, it is the conceit that status is a reward for talent and hard work. In health, it is the belief that "nature knows best." And in the academic study of evolution, the prejudice is echoed as an assumption that natural selection takes place only between individuals, never between teams or groups or communities. As these pages unfold, I hope to show you that these are prejudices, not science, and that the evolutionary prejudice is related historically to discredited economic and social ideas.

Aging Genes

Breeding animals is an ancient art. Dogs were domesticated long before recorded history. But in the 1980s, the breeder's art took a big step closer to being a quantitative science as techniques were developed to analyze small differences in DNA, to see exactly what genes were being bred. Tom Johnson, working at UC–Irvine, was onto the advantages of studying aging in the lab worm *Caenorhabditis elegans.* These worms are happy growing on a petri dish, they live just a couple of weeks, and their aging is remarkably flexible in response to feeding, temperature, and genes.

Johnson studied worms with a mutation in a gene that rendered it inoperative, as if it wasn't there. After he discovered that worms with the defective gene lived half again as long as normal worms, he dubbed the gene *AGE-1*. No one had ever imagined that a single gene could have such an effect on life span. In fact, the best experts in evolution had theorized that "everything ought to wear out at once," so that no single gene could have any noticeable effect. Johnson's discovery was the more remarkable because longer life required nothing new but rather the deletion of an existing gene. This implied that the effect of the *AGE-1* gene was to cut the worm's life short. What was it doing in the genome? How did it get there? And why did natural selection put up with it?

Johnson had a ready explanation. He believed (and still believes, I believe) in the conventional explanation for evolution of aging, answer number two in the section above (trade-offs and compromises). The worms without *AGE-1* laid only a quarter as many eggs as other worms. It was easy to see how they had been losers in Darwin's struggle. In fact, Johnson's finding looked like a dramatic confirmation of the theory that aging was a side effect of genes for greater fertility, greater individual fitness. Aging had not evolved directly, selected for its own sake, but as a cost of greater fertility, so the paradox was avoided.

But a few years later, this story unraveled, and what had been confirmation of theory number two became a direct contradiction. Johnson discovered that his mutant worms actually had *two* genes that were different. In addition to *AGE-1*, there was another, unrelated gene defect (*FER-15*) on a separate chromosome. By crossbreeding, he was able to separate the two. Worms with the *FER-15* mutation had impaired fertility without extended life spans. Worms with the *AGE-1* mutation had extended life spans with unimpaired fertility. This was a full-fledged Darwinian paradox: the *AGE-1* gene found in nature was the one that gave the worm a short life span. It was the "defective" gene that caused the worm to live longer. *AGE-1* looked not like a selfish gene but an aging gene. It was just the kind of gene that natural selection ought to eliminate handily. How had this gene survived, and what was it doing in the worm genome?

AGE-1 was only the first case of an aging gene in worms. There are now hundreds of genes known that lengthen life span when they are deleted.

In other words, these genes, when present, have the effect of shortening life span. Some of them tend to improve fertility; some don't. Some have other beneficial side effects, but about half the known life-shortening genes offer nothing in return, or at least nothing that has yet been identified. This is direct evidence in favor of answer number three, the answer that defies conventional evolutionary theory.

Aging genes have been discovered in other lab animals as well as worms. The other popular species for aging scientists to study have been yeast cells, fruit flies, and mice. These four species come from very different branches of the evolutionary tree. Nonetheless, they share common ancestors, going back hundreds of millions of years ago to the first eukaryotes (complex cells with nuclei and certain other organelles). You and I and the mouse and the worm and the fly and the yeast cell are all eukaryotes, and there are some genes—including deleterious ones—that we share. Why? Why has nature conserved life-shortening, "killer" genes?

The answer must reflect the unity of life. There are certain core functions of the cell that arose in the distant past and have been conserved through aeons of evolution. We all transcribe our genes into proteins using the same genetic code, we all burn sugar for energy using the Krebs cycle, and we all reproduce sexually using a style of cell combination and division known as meiosis. We all age and die.

It is remarkable that aging is one of these core life functions that almost all eukaryotes share. There are genes that modulate aging in yeast cells that are close cousins of aging genes in worms, in flies, and in mammals, including you and me. Despite the fact that aging is a disaster for the individual, evolution seems to have guarded and preserved the genes for aging as though they were the crown jewels. This is a dead giveaway that aging must have an essential biological function.

Patrician Genes

There is a natural analogy between population cycles in ecology and the economic cycles of boom and bust that appear in the brand of capitalism embraced by America and the West.

In the wake of the 1929 economic collapse, the Roosevelt administration was able to pass a broad program of government oversight and regulation of business practices. There followed forty years of unprecedented growth and the rise of an American middle class, the first time in history that any economic system had supported a majority of its citizens in comfort and security. But in 1980 began the Reagan backlash, deregulation, and the return of unbridled economic competition. Capitalism became predatory, the middle class began to shrink, business cycles deepened, and a growing rift now separates a wealthy elite from the struggling majority. There have been three major stock market crashes in the thirty years since deregulation, each followed by painful unemployment and economic stagnation.

Without regulation, competition becomes destructive. The idea that stability and a broad prosperity can emerge from pure competition is thoroughly discredited in practice, but it is a useful myth for the 1 percent who profit mightily when there are no rules and no regulators. If they told the truth about their motives, the rapacious corporate giants could never sell deregulation to an enlightened democracy. So they promote the dogma of a "free market," not because they believe in this or any ideology but because it supports the freedom of the largest and the strongest to pillage everyone else.

Historically, an important part of the argument for the benignity of free markets comes from the analogy with evolution. (Just look at the marvels nature hath wrought using only the chisel of bare-knuckled competition!) This is social Darwinism, the doctrine that the rich and successful not only contribute more to society than you and I but that they have better genes to boot. It is a perversion of Darwin's ideas, but from the very beginning, his sound biological theory has been linked to an elitist social ideology. During the early years of the twentieth century, social Darwinism played a crucial role in shaping the version of evolutionary theory that emerged and predominates to this day.

People are rich and powerful because their parents were rich and powerful—nothing remotely fair or just about it. But social Darwinism promotes the fiction that there is a natural order in the predominance of an elite hereditary class. Like the "invisible hand" of Adam Smith a century earlier, Darwin's "struggle for existence" has been caricatured to support the myth that pure, unrestrained competition can be magically transmuted into a harmonious society. The truth is that "tooth and claw" competition doesn't work in economics, and it doesn't work in ecology. Competition is vital, but it must be regulated, or it poses a fatal risk of instability for the whole system.

Darwin himself was of the British aristocracy, and he had some trouble acknowledging the huge evolutionary significance of cooperation. Yet in his later work *The Descent of Man*, Darwin spoke explicitly of a group that could have an evolutionary advantage distinct from the sum of its individuals.

By the mid-twentieth century, the existence of cooperation was denied utterly by the mainstream of evolutionary theory. We live now in an age of hyperindividualism, and it is no accident that, in our culture, the dominant version of Darwin's theory is based on pure selfishness.

Where This Book Is Headed

Aging is the very antithesis of what Darwin called "fitness"—the competitive vigor and reproductive potential of organisms. If aging is governed by genes that cause us to become frail, to lose our fertility, and to die, then how did those genes prevail in evolutionary competition? How could aging have evolved?

That is a central question in this book. The answer is that natural selection is not just the life and death of individuals but also the rise and collapse of local populations and of entire ecosystems. Evolution is about cooperation as well as selfishness, and aging has evolved as part of the dues owed by an individual for participating in a stable ecosystem. Evolution and ecology have inscribed a death sentence in our genes. We literally pay with our lives to protect ecosystems, as increased death rate via aging prevents the sort of wild overgrowth that presages wholesale population collapse.

Aging appears to be caused by an explicit genetic program. One consequence of this is that medical science will have to approach the diseases of old age with a different attitude. We can't use "natural medicine." We can't help the body to heal, because the body isn't trying to heal—it's trying to destroy itself. Instead, we must decode the signals that trigger self-destruction and replace them artificially with the signals of youthful vitality.

Another consequence is that the selfish gene version of evolutionary theory is not the whole story. Mothers teach their children to take care of themselves. What mother wants her child to be indifferent to the welfare of others? Mother Nature, like all mothers, has counseled her children against excessive selfishness. She wants them to take care of themselves to be sure, but because she wants them to get along in the world, she has tempered selfishness with altruism.

This seems to be an intuitive truth, easy to grasp for anyone who does not have an advanced degree in evolutionary biology (and to their credit, not a few professional evolutionists have figured out that the selfish gene version they learned in school was a tortured picture of reality).

Ripe for Scientific Revolution

The science of aging is a very active field, expanding in scope, with new labs, new techniques, and an influx of creative young talent. The field is not at all moribund, but it is flailing. At meetings and in journal articles, the puzzling mismatch between experimental results and theoretical expectations is everywhere apparent. Some researchers try to paper it over, and their explanations come out as incoherent or obviously flawed. But most are presenting the data with an honest admission, appending such comments as:

We don't understand why this happens.
There is some kind of dysregulation here, a failure of the body to do the right thing.
It's strange that the damage the body suffers seems completely avoidable, and yet . . .

What they are seeing—or refusing to see—is the body destroying itself on a schedule.

Evolutionary theorists have tried to understand aging within the framework of the theory as they learned it. They have done a workmanlike job of extending and modifying the theory as each new result comes in, carving out exceptions and elaborations on the basic themes. But specialists in the field tend toward myopia, and few have stepped back to look at the big picture, to see the signs that the basic principles of the theory no longer match the basic observations.

The essential problem is that the theory has been built from the bottom up, taking the individual animal and its individual success to be primary. (Darwin did not make this error; it came into the field only later, in the twentieth century.) Neglected are all the ways in which a community functions as a collective, with logic that cannot be understood simply as the sum of individuals. The limited perspective of the selfish gene is very much the standard today. Evolutionary scientists are quite aware that their paradigm is based on individuals to the exclusion of communal adaptation, and they have their reasons for this strong bias in favor of the individual over the group. Such reasons include an overreaction to an earlier excess in the opposite direction, when naturalists spoke too glibly about "the good of the species." But although the biomathematicians' skepticism of group selection was understandable, it is clear now that they are wrong. I* am in the minority, but I am not alone in saying this. Many smart people, including Nobel laureates and experts on the edge of the field, have recognized that evolutionary science today is missing something basic. Perhaps you have sensed this already; if not, I think that you will come to see it in the chapters that follow.

Why Should It Matter If Evolutionary Competition Is Individual or Communal? And What Does All This Have to Do with Aging?

Having a fixed life span—dying on a schedule—is bad for the individual, but it has advantages for the community. To understand what aging is and how it came to be, it is necessary to adopt a communal perspec-

* The pronoun *I* refers to Josh Mitteldorf, and *we* refers to both Josh Mitteldorf and Dorion Sagan (except where it is obvious from context that a broader "we" is intended).

tive for evolution. Aging is a paradox if your paradigm is confined to selfish genes, but it is possible to understand aging if you imagine natural selection in a broader context, with cooperative groups competing in a classical Darwinian struggle. The fashion in evolutionary theory these last fifty years has been to think exclusively in terms of individual competition, to disallow group competition. This has been the root of the scientific community's failure to understand aging.

I have said all this in scientific forums and on the pages of biology journals, and now I am going over the heads of the experts to appeal to your good sense. For eighteen years, I have been watching the response of the academic community to ideas about communal evolution, including my own ideas. It has been both deeply gratifying and intensely frustrating for me—gratifying because the scientific world is moving in the right direction, frustrating because the movement is so slow. There is still a great deal of unthinking bias against "group selection." Lab scientists still report their results in terms of the failed theories.

A few years ago, I was visiting my mother (now ninety-three years old and sharp as a tack) when I complained about the conservatism of the scientific community. I had wanted to initiate a scientific dialogue, and I was frustrated. "Take your case to the people," she said. "Write a book." That project proceeded in fits and starts for two years. Then I hit the jackpot and found an ally in Dorion Sagan. He quickly understood my ideas about aging and was able to help me both put them in the context of broader evidence-based evolutionary theory and make them more accessible. This book is our first collaboration.

PROLOGUE

Your Inner Stalker

· Dorion Sagan ·

The void, the concept of nothingness, is terrifying to most people on the planet. And I get anxiety attacks myself. I know the fear of that void. You have to learn to die before you die. You give up, surrender to the void, to nothingness.

—Harry Dean Stanton

When noir writer Jim Thompson titled his 1952 crime novel, *The Killer Inside Me*, he was referring to the unhinged mind of a small-town Texas cop, not the automatic behavior of cells. And yet we very well could have used his title on our book, for it turns out, you too are prey to an inner assassin. Unlike Thompson's cliché-spewing deputy sheriff, your killer has its sights set on itself.

The book you hold in your hands makes a case for something as astounding as it is unexpected: that our aging—which kills us if we don't die first by accident, war, or disease—is genetic. What is aging? Aging, or senescence, cannot merely be defined as the process of getting older. Time goes by for a rock, but it may not perceptibly change in thousands of years. To be more precise, we should define senescence—aging—not just as the passing of time but as *an increased chance of dying* with the passage of time. Thus, for example, my near namesake (my astronomer father might have been thinking about Orion, the constellation, closer to my spelling), Dorian, in Oscar Wilde's gothic classic *The Picture of Dorian Gray*, temporarily escapes the senescence that normally affects all of us after reaching adulthood. (See the Mortality Table and a more in-depth definition of aging on pages 58–59.) Dorian keeps his youth while a portrait of him, kept in a closet, ages instead. A nice arrangement—or a diabolical deal. After a failed relationship in which his actress fiancée kills herself with cyanide, Dorian becomes an unhappy hedonist, kills the painter of his portrait, bribes someone to dispose of the body, and finally, at his wit's end, stabs the now-hideous painting. At this point, the painting recovers its original form while Dorian ages eighteen years in minutes, his corpse identified only from his rings. Such is senescence, with a vengeance. And although it is fictional, such an "aging-in-a-hurry" lifestyle, although rare, seems to mark some bird species, such as the albatross. Senescence, as you'll see, is not a one-size-fits-all sentence of death by deterioration. It can be fast, slow, or—surprisingly—not at all. (See chapter 2.) These are the kind of raw, on-the-ground facts that mathematically hypertrophied but evidentially undernourished neo-Darwinism has had trouble making sense of. Until now.

I say that advisedly—very advisedly—because to say that suggests that my coauthor, Josh Mitteldorf, is a neo-Darwinist—a designation he stringently rejects.

Nonetheless, he shares with the best (and perhaps worst) of the neo-Darwinists a familiarity with, an attraction to, the development and analysis of mathematical equations and models that abstract the elements of biology in sometimes illuminating, even predictive ways. I am more philosophically inclined but have sometimes had the advantage, I would say, of being able to see the forest for not being distracted by the trees. Josh and I agree on this: it is, you might say, the underestimated advantage of the words out of the mouth of the not-yet-socialized babe; a kind of emperor-who-wore-no-clothes phenomenon. Like the child in Hans Christian Andersen's tale who stated the obvious—that the emperor whose gown everybody was pretending to admire was in fact without one—my glimpses of a scientific field were always from the outside. I did not get lost in the details because I did not know them, and I was not crushed by the inertia of grad school mantras I'd never heard. When, in first trying to popularize the idea of my biologist mother, Lynn Margulis, that evolution proceeds largely by symbiosis, I used the term "group selection"—to refer to the process of certain interspecies alliances and well-orchestrated societies outliving their less unified competition—she said, "You can't say that."

"Why?" I wanted to know.

"You just can't; the phrase isn't allowed."

In this book, you'll see why this was—and still largely is—the case. It is a core of Josh's fascinating new biophysical worldview, which has rather huge stakes for us humans, so long duped by the promisers of extended life and the religious counterpart, life after death. Josh himself, trained in astrophysics, comes to the field of biology, and specifically how organisms age, from the outside. But here he is in good company; the influential innovators of modern neo-Darwinism, from the right-wing Ronald Fisher to left-wing J. B. S. Haldane, came from the "hard" sciences, from physics, mathematics, and statistics. This was in fact part of the point: to put "soft" biology on a rock-hard foundation, even if that meant, ultimately, the sacrifice of common-sense assumptions upon the altar of mathematical workability and disciplinary prestige. When Richard Dawkins calls William D. Hamilton his "intellectual hero," he is paying homage not to the equations that attempted to demonstrate that aging is an evolutionary inevitability but to the curious, creative, but mathematically checked imagination that would devise such a thing.

Of interest to socialists and communists no less than capitalists and fascists, life does indeed seem to be a social phenomenon. Not just social but, as the accumulating evidence for *symbiogenesis*—the production of new life-forms from the melding of separately evolved organisms—suggests, hypersocial. The surface of this planet has been crowded for billions of years. And it is in this context of crowds, the benefit of their organization compared to the dangers of their rampant growth, that aging, according to Josh, must be understood.

Aging, in the view laid out here—which you can think of as a kind of giant case study of humanity's once-naïve view of the process—is not what you thought it was. This book does not contain your grandfather's view of aging. It does not contain the view of aging implied by my grandfather's advice, "Whatever you do, don't make the same mistake I made. About thirty years ago, I started getting old."

"Don't make the same *mistake*": clearly a joke, ironically underscoring that there is nothing we can do, wittingly or unwittingly, about an inexorable process. But according to the theory laid out here, my grandfather's assumption is dead wrong. Aging is no mistake. Rather, it is something that nature—however unconsciously—does *on purpose*. On the new view presented for the first time to a popular audience here, senescence does not come about through mere external exigency or as a result of the inevitable tendency for usable energy to be lost to the environment (*entropy*). Rather, as you will see, aging, or senescence, seems to be the result of multiple genetic systems all but conspiring to kill us from within. There are many signs that aging is genetic. The clearest proof is our ability to easily make organisms live longer. Fruit flies can be bred to live eight times as long. Starved rodents live 40 percent longer. Nematodes—tiny, fast-breeding roundworms, a favorite species for lab studies—insulated from the mere *smell* of food live significantly longer. The same is true of fruit flies. In other lab experiments with worms and yeast cells, geneticists have been able to increase life span just by disabling certain genes, which some have dubbed "gerontogenes." Josh refers to them more simply as aging or suicide genes.

Suicide genes present more than a little puzzle within the strict selfish gene paradigm. The realization that aging destroys organisms that have the tools to live longer—much longer—flies in the face of one of the most deeply held tenets of modern biology: that genes and individuals have evolved to maximize their own maintenance and reproduction. And yet there is now smoking-gun evidence that nature has been killing off her own for a good long time. Why? We know that Mother Nature is not always nice, but how could she—how could she do such a thing?

Josh shows us. Although aging certainly does not preserve individuals like you or me, it protects and invigorates communities, which, come to think of it, are integral parts of our species' survival. In evolutionary history, whole populations—not unlike certain economies and empires vulnerable in their dotage, too big *not* to fail—have routinely been completely wiped out. Suddenly running out of resources, their individuals starved to death. Or they became soft targets for predators that whittled their populations down beyond the point of renewal. Or they became sitting ducks for rapidly reproducing bacteria or viruses.

This population's-eye perspective has a checkered history. In its earlier phase, it was roundly ridiculed and in some cases appropriately criticized for careless rhetoric and naïve mathematics. But Josh, an innovator in biological computer modeling with a Ph.D. in theoretical physics, is no mathematical naïf. Evolutionary science is in the midst of a transition from a perspective based on pure competition to a broader schema in which competition and cooperation interact in complex and unexpected ways. If we expand our perspective to join such biologists as Lynn Margulis, David Sloan Wilson, E. O. Wilson (no relation), Stephen Jay Gould, Ernst Mayr, and Sewall Wright who see genes as only ever existing within cells and populations that compete at higher levels, we can begin to grok Josh's argument that aging—ensuring timely dying—doesn't "just happen" but is orchestrated by multiple genetic mechanisms.

Aging unto death you could say is a sexually transmitted disease, but the transmission is the normal way, through the parent genome. And the "disease," while you can bet good money it will take you out, is part of the health of the larger community. As Harry Dean Stanton quips above, there is a commanding nothing, a complete silencer, at the end of our road. But it may not be so much a cosmic conspiracy or a fluke of the universe that we are given the ultimate gift only to have it completely removed from us in the end. Rather, it may be part of a methodical evolutionary program. Nature may be the ultimate hard guy, with no moral qualms about keeping aging going, because it recycles matter and information, aiding in the production of its survival modes even as it protects against the dangers of overgrowth—choking pollution, food shortage and starvation, and becoming a wholesale meal, never to be consumed again, for the right predator.

The theorized adaptive role of aging protects against these potentially extinction-level events, as well as a virulent infection, which is likelier to sweep through a dense population of genetically similar organisms, killing them all, than to wipe out a group of organisms that die staccato, by aging, and therefore reshuffle their genes more quickly in sexually reproducing organisms. This aspect of Josh's idea overlaps with the famous Red Queen theories of sex.

Red Queen, the name, comes from *Through the Looking-Glass* and references a conversation with the Red Queen, an animated chess piece, who tells Alice she must run as fast as possible just to stay where she is:

the logic here mirrors that used to explain the persistence of sexual reproduction, which forces organisms to dilute their would-be selfish genes by hitching them to those of another being. The name Red Queen was first used by evolutionary biologist Leigh Van Valen, who applied it to species that must evolve—run as fast as they can—just in order to stave off extinction in an environment studded with other fast evolvers. Because Josh with John Pepper published a paper showing mathematically that the persistence of aging can be explained by its increase of variety—increasing sexual reproduction and genetic recombination by increasing the rate of turnover by death of individuals—I suggested he call his theory the Black Queen.

Formally similar to the Red Queen argument for the maintenance of sexually reproducing beings—they selflessly "give away" their genes because mixing them with those of the opposite sex ups their rate of recombination, keeping them a step ahead of predators—the Black Queen idea forces us to interpret the wholesale disposal of genes at death in terms of the positive effect it has in protecting populations. Like the Red Queen, which protects sexual reproducers from pathogens, the Black Queen increases genetic turnover, creating a more-difficult-to-hit "moving target." But there is really more to Josh's theory, and the name Black Queen is more shorthand, a kind of nonsuperstitious, scientific version of the more familiar Grim Reaper.

More than protecting them by keeping chugging the rate of group-protective genetic recombination, aging, or senescence, prevents populations from overshooting their resources, protecting them from total wipeout by starvation. This goes back to the aforementioned datum that smelling food can shorten your life—that is, if you're a nematode or fruit fly. The widely noted connection between not eating too much—caloric or dietary restriction—and longevity also makes sense here: food in the environment is not just a nutritional resource but a signal in the environment that there is no danger of starvation—at least not yet, but there may well be if the population grows too fast, destroying its resource base. The essence of Josh's idea is that too quickly booming populations are more likely to bust, and this has happened so many times that evolution favors those that adjust their population size so as not to be a flash in the evolutionary pan. Paradoxically, death—genetically doled out by internal poisons like free radicals* or by restricting the body's access to needed enzymes like telomerase—kills us but buttresses the groups of which we are part, our genetically connected populations.

Although group selection has been badly maligned, the old glacier of instant dismissal of the importance of groups in evolution is beginning to thaw. Daniel Dennett, for example, in his afterword to a revised edition of Richard Dawkins's *The Extended Phenotype*, writes:

> *Each of us walks around each day carrying the DNA of several* thousand *lineages (our parasites, our intestinal flora) in addition to our nuclear (and mitochondrial) DNA, and all these genomes get along pretty well under most circumstances. They are all in the same boat after all [like] an individual human being, with its trillion-plus cells, each a descendant of the ma-cell and pa-cell union that started the group's voyage . . . [quoting Dawkins] "At any level, if a vehicle is destroyed, all the replicators inside it will be destroyed. Natural selection will therefore . . . favour replicators that cause their vehicles to resist being destroyed. In principle this could apply to groups of organisms as well as to single organisms, for if a group is destroyed all the genes inside it are destroyed too." So are genes all that matter? Not at all.*

In the following pages, Josh—with some minor help from me—embarks on a meticulous examination of evolutionary concepts of aging, showing how what Dennett calls "groupish entities . . . [a] herd of antelope, a termite colony, a mating pair of birds and their clutch of eggs, a human society" are routinely destroyed when they don't evolve protective measures against ecologically perishing, paradoxically, from overpopulation. Leigh Van Valen's famous Red Queen Hypothesis of the

* Free radicals are unstable molecules that can oxidize neighboring tissue, destroying it. They are implicated in aging via programmed cell death but also may be helpful—for example, when deployed by the immune system against cancer.

1970s holds that organisms *must* evolve not in order to conquer but just to keep up with other organisms that are evolving. Change or die. As said, the most famous application is to sexually reproducing organisms, whose genetic combinations via sex present a more difficult, moving target for microbial pathogens that quickly reproduce and evolve. If this is right, then our sexual desire may in part reflect the success of macroscopic organisms recombining faster than microscopic infectors can kill them. It even may be interpreted to suggest that the desire for variety in sexual partners is a human manifestation of an animal (and plant) adaptation for staying alive—not at the level just of the individual but also of the population and species. Black Queen logic is similar. Like the wandering eye of the Don Juan providing an advantage to his gene-mixing population, death cleans the slate, making way for new, potentially more resistant genetic combinations, improving the chances of populations to achieve maximal genetic diversity. It also, in the regulated release of aging, protects the population from wiping itself out in the aftermath of overgrowth—a real concern now for *Homo sapiens*, a species who seems to have used its "intelligence" to circumvent this long-evolved mechanism.

Josh's new thanatology represents a fascinating biological breakthrough that we must heed carefully if we are not to add to the overpopulation that brought ever-evolving life into the business of genetically programmed dying in the first place. If we were all to intentionally trick our bodies into living longer than we've been optimized to live—or if modern medicine finds a wholesale way to suppress our suicide genes—the results could be catastrophic for the fate of our species as a whole. Such a wholesale implementation of just the sort of individual selfishness that evolution was trying to avoid could spell doom—winning the aging battle and losing the survival war. I note that at present rates of doubling every half century, we will have *quadrillions* of people by the year AD 4000. Our technical intelligence may have gotten us into this bind but may not get us out. Yet similar problems have, however, been dealt with in evolution before, by just the sort of group selection that aging epitomizes. And there is a philosophical angle. Death is a harrowing prospect, as horrifying as it is inevitable. Yet on the views herein, each of our aging-unto-death, each of our declines, can be seen through a more accepting lens. Rather than fight aging, one can be comforted by it, viewing it as a way that many species with a tendency to grow too fast have managed to survive, as ours has, so far, and perhaps still will, its members changing and aging, being reborn and recombined, into the distant future.

Josh's work thus helps us understand the strange naturalness of our deaths, and there could hardly be a higher encomium. Fear not the Reaper—even if he is lurking from the inside. Nature knows what she's doing. In fact, she's a bit of a dominatrix. It's a little jarring at first, but ultimately, I don't mind so much. Better the devil you know, as they say.

Before I go, let me say that I am an avid advocate of Josh's evolutionary logistics. They are part of a new program of health regime that makes better evolutionary sense than any I've yet seen. Attempts to cheat aging are legion, and we should expect them to be bogus. The Spanish conquistadors who found America sailed in search not just of gold but of the fountain of youth. Ponce de León was still hunting for such a natural source of eternity's elixir when he was shot dead with a poisoned arrow in 1521 at age forty-seven. Centuries later, the Italian mystic Alessandro di Cagliostro promoted an elixir that he claimed would stretch a life span to three hundred years. Cagliostro died in 1795, aged fifty-two. I'm no Cagliostro or Count of Saint Germain—his supposed mentor, chronicled as existing in several centuries, for example, in Casanova's diaries—and I'm critical when I'm told what I want to hear. In "Dr. Heidegger's Experiment" by Nathaniel Hawthorne, the eponymous host invites his hoary guests to imbibe a rare potion. After a few sips, they find themselves charmed, alive, flirting and dancing and marveling at their images in the mirror. But alas, the effect expires at the close of their dinner party, and they realize the drug was a hallucinogen rather than a youth serum: it only made them dream they had reversed aging.

There are no miracles in this book. But Josh's work has driven home the point that aging doesn't "just happen." That implies the possibility of a whole new strategy for thwarting it. We don't have to fix everything that goes wrong if we can only trick the body into thinking it is younger.

This in fact is already being accomplished—in baffling ways that don't make sense in terms of previous theory.

When we notice, with the help of this book, that some species (clams, sharks, turtles, lobsters) show no signs of aging in the precise statistical sense of being more likely to die next year than this year, the notion of cheating aging, or "tricking the Black Queen," slips outside the grasp of the con artist and drifts away from the realm of the literary dream. It moves squarely into the land of reality. Strangely, as you'll see, eating plenty of nutritional foods or eating too well actually takes years off our lives. And yet the body can be tricked into thinking it's not getting enough without you having to go hungry. Excited both about the knowledge itself and the opportunity to put it into practice, I have begun—as seems certain will many of the readers of this work—to apply these evolutionary logistics to protect my own life, health, productivity, and longevity. Vitamin D, Kundalini and ketogenics, melatonin and meditation, intermittent fasting and calorie restriction, basketball, walking and dancing, alternating hot-and-cold showers (part of shiatsu and a Black Queen protocol), red wine, and sunshine—the recognition of the Black Queen adds to our self-knowledge and the effort to "cheat" her to live longer also gives us more time and a greater feeling of well-being. Understanding that nature is not going to protect you in pristine condition just because you take care of yourself—in fact, she'll take care of you in the Mafia, hit man sense of the phrase—is a wake-up call in multiple areas.

INTRODUCTION

How a Lifelong Obsession with Aging and Health Became My Career

· Josh Mitteldorf ·

Fear of Death, and Fear of Fear of Death

I was three years old when my father told me that someday I was going to die. I was terrified. The thought that a few decades of life would be followed by an eternity of nothingness obsessed me, drove me frequently into a panic, and sent me crawling into my parents' bed in the middle of the night. I know now that this kind of fear is not uncommon in young children, but attaching it to something so abstract and distant seems unusual.

I had immediate, firsthand experience with the crippling incoherence that naked fear could evoke. No one had to tell me that fear itself—not death but the fear of death—was a horrible, unbearable plague. But as a small child, I was silenced by shame and embarrassment. I presumed that susceptibility to fear was my peculiar weakness and that I was all alone in having to come to terms with it. I learned to distract myself and put thoughts of death out of my head. I told myself that someday I would face the specter of death, but for now, it was just too uncomfortable. The bargain that I made was that I would allow myself the luxury of distraction with the promise that I would return to this issue of mortality and sort it out when I was thirty-five. Yes, even as a preschooler, I had a certain affinity for numbers, and I thought thirty-five was safely tucked away in the remote future but still only halfway to the age my father told me I might expect to live.

As it turned out, I was off by about a decade. At age thirty-five, I was delightfully occupied with the adoption of my first daughter, but when I was forty-six, a confluence of inner readiness and outer events drew me into the contemplation of death. It began with scientific study, which led not only to other studies and this book but also to a more youthful body, renewed energy, better health, and a feeling of relaxation and empowerment in an area that had once paralyzed me with fear.

Cancer Concerns and Pollution Paranoia

I came of age in the 1960s, just as the term "natural" arrived on the scene. I thought then, as most people still believe, that staying healthy and improving my odds for a long life were the same thing. My ideas about how to stay healthy were to give my body all the things it needed: vitamins, minerals, and complete protein, plenty of rest, moderate exercise, and a low-stress lifestyle. I aimed to sleep nine hours a night for the same reason that I aimed for 120 grams of protein—that's a pound of lean meat every day—because more is better.

I was afraid. Most of all, I feared cancer. Any tiny dose of radiation, any food additive or pesticide or pollutant in the air might trigger a carcinogenic mutation. I now think of cancer as a systemic disease, but at the time, my belief was that a single rogue cell, just one unlucky break, could spread to kill me at any time. That belief in itself was a recipe for paranoia. The idea that I was being poisoned by modern life supplied a target for my obsession. Air pollution made me nervous, and cigarette smoke drove me to distraction. This was the 1970s, and cigarettes were ubiquitous, even in California.

I was an astrophysics student at UC–Berkeley, using computer models to study the cosmos. Though I was a scientist by temperament as well as profession, it would be years before it occurred to me to look into the science of aging or even to learn what medical science had to say about the lifestyle correlates of longevity.

I ate crunchy granola and whole-grain bread. I tried nutritional yeast and lecithin and spirulina and became enthusiastic with each new health and longevity miracle I read about. Vegetarianism was still confined to a fringe of health nuts and Seventh-Day Adventists. When I began yoga in 1972, I think Berkeley was one of the few places in the country where you could find a weekly yoga class. Over the years, yoga would train me in sensitivity to my own body that provided an experiential knowledge stream that I now think of as complementary to clinical data.

One evening, about six months into my discovery of yoga, I was lying on the floor in *savasana* (deep relaxation—literally "corpse pose") when the voice of my revered and beloved teacher suggested to the class that perhaps we might find our practice leading to eating less meat. I was startled awake and sat bolt upright. In previous weeks, she had suggested cutting back coffee and alcohol and TV and marijuana (this was Berkeley) and cigarettes—it all went down smoothly because I had never been attracted to any of those things. But what could she be thinking, lumping meat with intoxicants and mind-altering drugs? I had never questioned that a diet that was ultrahigh in protein might not keep me strong and healthy. The phrase "new age hokum" hadn't been invented yet, but those are just the words for which my mind was fumbling.

Six weeks later, I was a vegetarian, and I have never looked back. My teacher's hypnotic suggestion awakened my latent discomfort with the killing of animals. It had nothing to do with science. Now there is evidence linking low meat consumption with longevity, but I certainly didn't know of any at that time.

It was 1982 when I made friends with Howie Frumkin (now dean of the University of Washington School of Public Health). Even at that time, fresh out of med school, his easy warmth and twinkling eyes coexisted naturally with a commanding intellect. I saw him in his office at the Hospital of the University of Pennsylvania and confessed that I had been losing sleep over worries about cancer for as long as I could remember. "Cancer is a disease of old age," he told me. He sat me down and showed me the charts. With the exception of childhood leukemia, most forms of cancer were a very low risk for young people, rising steeply and peaking between the ages of seventy and ninety. This was completely new to me and very welcome to hear. I was relieved of an obsession.

Alfalfa and Aflatoxin

In the mid-1980s, Bruce Ames launched another seismic shift in my longevity program with a series of articles in *Science* magazine about natural pesticides. Ames had been studying carcinogens in the diet and had come to prominence with the invention of the Ames Test, a quick way to screen food additives for carcinogenic potential that has saved billions of dollars for the industry and obviated the slaughter of thousands of innocent rabbits.

I was the stereotypical Mr. Natural, based on my belief that pesticides and preservatives in my food were the biggest threats to my life and health. Along came Ames with a new story. It seems that humans didn't invent insecticides. For as long as there have been beetles and grasshoppers on the planet, plants have been manufacturing chemical weapons to protect themselves. Some of these pesticides have been found to be carcinogenic in tests on mice and rats. But according to principles of the Food and Drug Administration (FDA), they cannot be banned or regulated or even labeled. They come under the category GRAS—"Generally Recognized as Safe."

For many years after the Ames Bomb, I annoyed and inconvenienced my family (my wife was most patient) by refusing to eat black pepper, beets, alfalfa sprouts, peanut butter (aflatoxin), parsnips, potatoes (solanine), basil, celery, mustard, and spinach (oxalic acid). These items topped Ames's ranking of hazards in the American diet, based on a combination of lab tests for carcinogenicity and prevalence in our foods.

Broccoli was on that list too . . . but how could I give up broccoli?

In the spring of 2014, a distant relative appeared from nowhere and e-mailed me with a family tree on the side of my paternal grandmother. She told me that Bruce Ames is my second cousin once removed. I was charmed. At eighty-five, Bruce still runs an active lab at UC–Berkeley, his eyes twinkle more than ever, and he continues to publish innovative research.

I will always have the deepest respect for Ames and his work, but I no longer place so much importance on his approach to toxins in the diet. A modest load of toxins in the diet is actually good for us, and we're likely to live longer with the toxins than without. We'll return to this counterintuitive idea—opening onto one of the main counterintuitive messages of this book—in chapter 6.

Epigenetics and an Epiphany

In January 1996, I read a *Scientific American* article about caloric restriction and life extension. Professor Richard Weindruch, a biologist from the University of Wisconsin, told of his research with animals that lived longer the less they were fed. It wasn't just a quirk of a lab rat's metabolism. The experiments included dogs and spiders, yeast cells and lizards, and now Weindruch was working with rhesus monkeys. They all lived longer on a starvation diet.

It was this revelation that began the shift in thinking that has carried me to my current understanding of aging, its evolutionary origin, and its deep relationship to health. Within a few days of reading this article, going for long walks in the park, scratching my head, I knew I had been battling the wrong enemy. Aging is an inside job, a process of self-destruction. I drew this message from the fact that the body is able to forestall aging when it is in extreme deprivation, desperately slashing its energy budget to conserve every calorie. This means that when food is plentiful, aging is avoidable, but the body is not trying to avoid it. Aging, it seemed, must be programmed into our genes.

You can call it a lucky guess, or perhaps it was a big-picture perspective that only an outsider in the field would have. The insight that aging is programmed into our genes has been at the center of my research ever since, and it is also the principal theme of this book. Unknown to me, there was other evidence for this insight even in 1996. There is much more such evidence today. Some of the genes that regulate aging have been discovered, and some of the epigenetic mechanisms that make aging happen have begun to come out. (Epigenetics is the science of how genes are turned on and off.)

This theoretical insight came with a bonus in the form of practical implications for self-care. All that good whole-wheat bread and organic tofu had begun to make its mark, and my midriff had the beginning of a spare tire for the first time in my life. I had been fortunate to have one of those metabolisms that let me pack away food without consequences, but now my weight was about ten pounds over what it had been in my twenties and thirties. I immediately began to cut back, to lose weight by brute willpower. It was harder than I'd thought it would be to lose the weight, but it felt great. I had so much energy that I took up running, and I did my first half marathon that fall. I also experienced, at last, a loosening of the grip in which fear of death had held me since childhood.

I was learning that I had been looking for longevity in all the wrong places. My emphasis on maximizing nutrition and minimizing toxins had been misguided. I had missed a great truth about keeping healthy, but more than this, I had misunderstood the nature of the enemy. All my thinking had been rooted in vague, unformed ideas about what aging was and how it worked. For me, science, health, and aging were beginning to come together for the first time.

The health message was surprising and disorienting, but there was another thread to this story that tickled my intellectual interest. I wondered about the caloric restriction effect (CR) and how it might have evolved. Every function of every cell and organ in our bodies has been shaped by a process of evolution and can only be understood in that context. How could life extension be an adaptive response to starvation?

Many different animal species respond to CR. This can only mean that there is some very general value in living longer when food is scarce. If evolution has produced this same expedient for so many different species, then it must have a purpose, and that purpose must be so general that it applies to yeast cells and to dogs.

But what could that purpose be? I guessed the reason why starving animals should have access to extra strength: it must be to help them survive through a famine. It was still unclear to me why aging would be programmed into our genes, but for some reason, nature prefers a fixed, predictable life span to a life span that is subject to the viccisitudes of chance. If natural selection was favoring a length of life that is not too long or too short, then whenever there's a famine, it makes sense that aging would get out of the way, since so many lives are already being cut short by starvation. Conversely, aging has to take a big bite out of the life span in times of plenty; otherwise, there would be no room to expand the life span when conditions changed.

From an Age Before Spam

These ideas about caloric restriction and the evolutionary origin of aging were fascinating to me—indeed, the most intellectual excitement I had experienced since encountering the big ideas of cosmology a decade earlier. I put together an essay—concise, naïve, a bit self-important—and sent it to an e-mail list of about a thousand evolutionary biologists that I had found online. Now, this was a time when the World Wide Web was young and text-based. E-mail had begun to expand from government and universities to more general use, but there was as yet no spam. Can you remember a time when the Internet was pristine? There was a gentlemen's agreement that, even though bulk e-mailing was essentially free, we would not permit the Internet to be polluted with unsolicited commercial messages. So my message wasn't discarded or ignored.

I received about thirty replies, some of them very generous and solicitous. All of them told me that my thinking was in error, because evolution doesn't work for the community but only for the individual. People took the time to explain to me that I was making a common error, one that other scientists had made in the past, but evolutionists had corrected their thinking in the 1970s. There is no such thing as "group selection." Natural selection works solely for the benefit of the individual. They told me to read *Adaptation and Natural Selection* by George C. Williams.

So here was a genuine scientific mystery dropped into my lap. I rapidly began to take ownership of it and more gradually gave myself over to its exploration.

I had imagined that evolution might arrange for animals to hold some strength in reserve for hard times. How did this require "group selection"? And if it did, what was so unscientific about "group selection"? I had a lot to learn. I was not so arrogant as to doubt that these experts who had graciously answered my question might know something I didn't know. But I was curious that none of them offered an alternative explanation for the evident paradox: that starving animals are able to live longer, healthier lives than those who have all the nutrition they need. I resolved to keep an open mind about whether my thinking had been flawed in the way that these experts suggested, or whether perhaps I was seeing something the experts had missed.

The Experts Were Right

As a scientist, I'm more of a thinker than a reader. Faced with a new problem, I'm inclined to go for a long walk and allow my thoughts to sift or to scratch equations in a notebook or even to try a stripped-down example represented by numbers in a spreadsheet. Compared with Googling the answer, this process is terribly inefficient. It also leads me astray, and I get things wrong at least as often as I get them right. I continue in this way first because there is no satisfaction so sweet for me as engaging with a scientific puzzle. I rationalize the inefficient use of time with the hope that trying lots of wrong ideas and following them to the end gives conviction to my knowledge and a depth to my understanding of how the world works.

But after many walks in the park, I realized the experts were right to say that if evolution has a preference for aging, this would have to be via group selection. If Darwinian competition were between individuals that live a longer time and individuals that live a shorter time, the ones that live a long time would leave more offspring, and their genes would come to crowd out the short-lived genes. This is not to say that aging cannot evolve—there is still the possibility that a community of individuals with a fixed life span is better adapted in some ways than a similar community where life spans are allowed to vary all over the map. For this advantage to become a factor in evolution would require competition of one community against another, and that is what the experts meant by "group selection." (You don't have to understand this now—I certainly took long enough to appreciate it. It will become clear bit by bit.)

But all the walks in the woods that I had enjoyed that summer could not elucidate for me, what was the objection to group selection? Why were the experts so convinced that group selection was not part of evolution's toolbox? After all that thinking through things on my own, I finally read the book by Williams that so many evolutionists had recommended. I found it stimulating and thought provoking. It opened my eyes to a more concrete and disciplined way of thinking about evolution. But I still didn't understand what was wrong with group selection. Could this be just a scientific prejudice among all these experts?

Old Darwin and New Darwin

I affiliated myself with the Biology Department at the University of Pennsylvania because it was convenient and close to home. I talked to professors, took courses in evolution, and read books to learn how evolutionary biologists think. I learned that for the last seventy years, the field has been dominated by a methodology known as "population genetics" or the "new synthesis," or a third name—the one I'll use throughout this book—"neo-Darwinism." Neo-Darwinism is not the same thing as Darwinian evolution. Darwin was a naturalist, a student of the natural world who described what he saw and tried to integrate his observations with understanding. His thinking was (appropriately, I think) vague and even modestly self-contradictory at times—his enthusiastic contemporary Samuel Butler even compared it to the "twitching of a dog's nose"—as he sniffed out the various ways that natural selection can work. Neo-Darwinism arose in the 1930s, and it was an attempt to make Darwinian theory more quantitative and rigorous. In fact, the field was founded by mathematicians who knew little about actual biology. As a physicist, I found myself immediately comfortable with the style and the methods of neo-Darwinism. It is straightforward and logically compelling. But the more I became immersed in the theory, the more I found that neo-Darwinism doesn't work very well as a description of real life. Several big things about life in general just don't add up in the context of neo-Darwinism: There's aging and death—I'll try to show you in the coming chapters why I don't think you can account for the basic facts about aging within the framework of neo-Darwinism. But in addition, neo-Darwinism can't account for sexual reproduction or for the structure of the genome that seems actually "designed" to make evolution possible; neo-Darwinism also does not have a place for the recently established phenomena of epigenetic inheritance or horizontal gene transfer.

One day in the Biomedical Library at Penn, I looked up a paper by one of the most respected theorists in evolutionary science, John Maynard Smith. The title of the paper was "Group Selection." Perhaps Maynard Smith's writing was more lucid than I had encountered before, or perhaps I was finally paying attention and giving the authorities their due. In one of those moments when the vase becomes two faces, I understood why the best theorists in the field had rejected the idea that natural selection could act on groups.

Evolutionary novelty depends on mutations that arise by chance. The mutation appears first in a single individual, and from there, it either spreads through the population or dies like the proverbial flash in the pan. Suppose a mutation arises that is bad for the success of the individual but would ultimately be good for the community if everyone should adopt the trait. A tendency toward cooperation is a good example. It does nothing for the individual and in fact will probably put the individual at a disadvantage if she is the only one cooperating while everyone else is taking advantage of her help but offering nothing in return. Of course, a cooperating community can be vastly more effective at group tasks than a community in which every individual is for himself alone. But how do we get there? The gene for cooperation started out in just one individual, and there is no reason to suppose it can spread to dominate the group. In fact, natural selection is working against it from the start. If such a gene can't spread through the group—an uphill battle—then we'll never find out whether it benefits the group or not.

I understood for the first time why the experts were dubious about group selection. I remember feeling queasy in my stomach as I bicycled home from the library.

A Worthy Scientific Puzzle

So I was not to have an easy victory or even the smug satisfaction of knowing that I had seen in an instant what evolutionary experts failed to see. But over the ensuing days, I came to realize that what I had was a genuine conundrum, a worthy scientific puzzle. It was just as clear to me as ever that evolution had chosen to cut life spans short, had installed aging in the genomes of the great majority of animal species. But I now appreciated the paradox: aging is bad for the individual, good for the community. How did the genes that control aging manage to spread and take over any community, a prerequisite for their communal advantage to become effective? How has aging managed to persist in the genome when any rogue individual might mutate away his aging genes and become new ruler of the roost? This was a puzzle with which I might engage deeply and find challenge and satisfaction. With some luck, I would be able to transfer the skills in mathematical modeling that had been so useful in my physics career and apply them in a new area.

That same year, I came across a feature in the Tuesday "Science" section of *The New York Times* about a professor at Binghamton University who had devoted his career to the study of group selection. It began,

> DAVID SLOAN WILSON *was a newly minted Ph.D. in his early 20s when he went to visit one of evolutionary biology's leading theorists and tried the intellectual equivalent of selling atheism to the Pope.*
>
> *"I just walked into his office and said, 'I'm going to convince you about group selection,'" Dr. Wilson recalls. He failed. His target, George C. Williams of the State University of New York at Stony Brook, had made his reputation by effectively wiping that very idea off the intellectual map only a few years before, in a 1966 book,* Adaptation and Natural Selection.

I wrote to David, and he was gracious enough to invite me to drive up and spend an afternoon with him at Binghamton. In the ensuing months, we collaborated closely together, and it felt like a homecoming for me. Over time, he introduced me to the friendly, cooperative left wing of the community of evolutionary biologists—those who study and advocate the process of group selection. I had a fascinating problem to investigate. I had a mentor. And I had a foot in the door.

ONE

You Are Not a Car: Your Body Does Not "Wear Out"

Just as the constant increase of entropy is the basic law of the universe, so it is the basic law of life to be ever more highly structured and to struggle against entropy.

—Václav Havel

I'm an extraordinary machine.

—Fiona Apple

Doesn't it seem strange that our bodies build themselves miraculously from single egg cells to fully formed, perfectly functioning adults—but then they can't seem to maintain themselves in good repair, as they gradually deteriorate and ultimately fail? It's as if the teen next door, who pieced together a complete 2002 Toyota Camry out of scrap from a junkyard, ran the car into a pothole and couldn't figure out how to change a tire.

Surely it is not a question of *can't* but *won't.* The body knows perfectly well how to repair and maintain itself, but that is not part of its genetic program, not part of its evolutionary mission. Modes of repair are shut down progressively as we age.

How Do You Think About Aging As We Begin?

Before I speak to people about aging, I always ask what ideas they have come to on their own. Everyone has thought about aging, at least enough to come to terms with it in their loved ones and their own lives. What is aging, and where does it come from?

When I am speaking to an audience of evolutionary biologists, the majority answer that "aging is a pleiotropic side effect of genes for fertility." This is what we referred to as "answer number two" to the mystery of aging in the preface. But outside university evolution departments, I have never found people to respond in this way. Instead, there are two popular notions about aging. About half the educated public already has the right idea (as I shall argue) about the significance of aging; this chapter is addressed to the other half.

The right idea is that aging has been programmed into our genes by evolution, an adaptation to make room in the niche for the next generation to grow up. The effect is to democratize, to keep the community diverse and resilient, and above all to stabilize the ecosystem against lopsided growth of any one species. The wrong idea is that bodies wear out for the same reason that machines do—gradually rusting, accumulating little nicks and dents. If that is the way you think about aging, my goal in this chapter is to convince you otherwise.

Things wear out. Nothing lasts forever. This is the oldest and still the most pervasive idea about what aging is. It is seductive because some aspects of aging fit with this picture; but the idea is also deeply flawed. It is a misapplication of basic physical law, and it also fails to account for some familiar facts about aging.

Bodies Versus Machines

The joints and bearings in your car become pitted and rusted over time, and they continue to work, but with less freedom and more squeaking. Isn't that just what happens to our arthritic knees and shoulders as cartilage wears away? Knives get dull, and the blades develop nicks and chips—just like our teeth. The plumbing pipes in your house become corroded over many decades, and deposits build up on the interior walls, impeding the flow. This sounds a lot like atherosclerosis—coronary heart disease. Often the performance of an older automobile engine suffers because the piston rings wear away, leaking exhaust within the cylinders. Our athletic performance declines with age, and it is natural to imagine it is for similar reasons. Biochemists might speak of "leaks in the electron transport chain" of our mitochondria, which are mini power plants inside each cell. In snowy, salty winters of the American Northeast, car chassis rust out over the years; as New Englanders might say, "The car has cancer." Even computers that have no moving parts are subject to performance degradation with age, because more and more apps are running in the background, each grabbing a chunk of the processor's "attention." Our immune systems fail in a similar mode, as our bloodstream accumulates *memory T cells,* white blood corpuscles that are expert at responding to diseases from our past, but there is a shortage meanwhile of *naïve T cells* that can respond to the next challenge.

Most compelling of all are the big problems that appear suddenly and send your car into the shop. The transmission fails, or the brakes wear out, or a rusting exhaust line finally leaks through into the passenger compartment. These problems are much more frequent in an old car than in a new car, and they are the reason that we retire the old car and buy a new one. Our bodies, similarly, become more vulnerable with age. Your chances of suffering a heart attack are fifteen times higher at age eighty than at age forty. You are twenty times more likely to be diagnosed with cancer at age eighty, compared to forty (and ten times more likely to die of cancer). Incidence of ordinary infectious diseases also rises with age, and though the increase is not so dramatic, the same diseases can be far more serious in an elderly person. Taken together, pneumonia and influenza are the eighth leading cause of death in the United States, and almost all these deaths are in people over seventy, most over eighty.

These surface similarities mask some major differences. If you leave your car in the garage most of the time and drive only two thousand miles a year, it will last a lot longer. But if you stay at home and don't use your muscles, you are risking rapid aging and early death. Exercise is the best thing you can do to extend your life. Why doesn't exercise wear your body out the way fast driving wears out a car?

This is our first clue that there is something very different about aging in a living body compared to wear in a machine. The body can fix itself in a way that the machine cannot. So for the body, its state of repair depends on the difference between the damage that is done in living and the repair that is accomplished internally, automatically, by grace of complex evolved physiology.

Aliens, observing Earth from space, might conclude that cars, with their hard exteriors, were exoskeletal life-forms, similar to giant beetles with rotary feet. They might even think the soft-bodied beings entering them were some kind of strange symbiont. But closer inspection would reveal that cars don't metabolize. They break down fuel to move, but they don't use it to continuously maintain themselves. A car doesn't grow (let alone regrow) its tires, its steering wheel, its hood, its engine. A car doesn't deploy Rust-Oleum to fix its fenders or destroy itself with energetic tools that might help it. A living being does many such things.

Better Than New

You might imagine that the body is *always* doing its best to repair itself, subject to some limits imposed by an energy budget. If the injuries and the damage are at a manageable level, then the body's automatic repair service will keep up, but if you are hard on your body, the repair falls behind, and damage accumulates. That would be a reasonable expectation, but think about it—that's not at all the way your body actually behaves. If you sit all day, get no exercise at all, nothing that would stress the bones or tear the muscles, then the muscles will atrophy, and the bones will soften. On the other hand, if you run or jump or lift weights, then the bones develop microscopic cracks, and the muscles will have tiny tears. Yet the body responds to these stressors by coming back stronger: in this case by building new bone mass, strengthening the muscles.

The more you work your body, the more the repair mechanisms are ramped up. That's not surprising—it's just what you would expect if evolution acted as a systems engineer might, allocating resources where they can do the most good. But the strange thing is that the body *over-compensates*. The body that works hard stays in very good repair and lasts longest. The couch-potato body is not stressed at all, yet it suffers rapid damage and dies young.

This is not the way a systems engineer would design our mortal coils. A rational designer would budget more resources for repair when the body is working out and needs it more, but it would not neglect the body and let it go to pot when repair is easy. The paradox is that just when resources (i.e., food) are most available and the need for repair (from exercise) is minimal, the body does a lousy job of repair, inviting early death.

Food energy is the currency of the body, the raw material with which the body must maintain itself, compete in the struggle for existence, and reproduce. A well-accepted theory of aging (the Disposable Soma Theory described in chapter 4) is based on the assumption that animal bodies rationally allocate the limited resource of foods. If strapped for resources, your body would be expected to address the most immediate priorities first and skimp on the long-term investment. The immediate priority is survival and reproduction; the long-term investment is the healing that keeps the body in good repair. The theory states, quite reasonably, that the body ought to do the best job of repair when there is plenty of food and when other demands for that food are smallest. So the Disposable Soma Theory predicts that we ought to live long when we eat a lot, exercise very little, and have no children. Women ought to live much shorter lives than men, because they invest so much more energy in reproduction than men do.

The truth is the opposite on all counts. Women have longer life expectancies than men. (In most animal species, too, the females outlive the males.) And women who have more children have a slightly longer life expectancy. The more you exercise, the longer you live. And food deprivation is a royal road to a long and healthy life.

The best-studied, most reliable laboratory manipulation for extending an animal's life is to feed it less. The less an animal eats, the longer it lives (on average). Animals that are completely emaciated, on the brink of starvation, live longest of all. Mice that are starving and running (miles per day!) on their tread wheels live even longer than the starved mice that don't exercise.

Why should life be cut short when food energy is most plentiful and competing demands for that energy are least severe? We must confront the puzzle that the body is *not* doing its best to live as long as possible under these circumstances. Damage is accumulating most when it is least necessary. In this sense, aging seems gratuitous, if not perverse.

In the original *Star Trek* episode "Shore Leave" (written by science fiction great Theodore Sturgeon) Mr. Spock* responds to the suggestion that he take some time off and join the crew in running about and frolicking on a vacation planet. Spock says, "Not necessary in my case, Captain. On my planet, to rest is to rest, to cease using energy. To me, it is quite illogical to run up and down on green grass using energy instead of saving it." A fact so familiar we don't think about it, but one nonetheless exceeding strange is that exercise increases health and extends life span. Spock's attitude is indeed logical but not empirical for an organism, where exercise increases strength and can increase both health and life span.

*** Played by Leonard Nimoy (1931–2015) whose last tweet was "LLAP"—"live long and prosper"—the saying (coined by Sturgeon) of his Vulcan character.**

Plants and Animals That Don't Age

It's a curious fact that some animals (and many plants) don't age at all in the technical sense that they become more likely to die as they grow older. This is another set of facts that doesn't fit with the idea that aging is an inevitable result of wear and tear. There are clams and lobsters that just grow bigger every year, without any of the body part failures associated with old age. There are groves of aspen trees cloned from a single root that are more than ten thousand years old. There are salmon and octopuses and seventeen-year cicadas that have extended periods of development in which they suffer no aging at all; but once they reproduce, they then age and die in a hurry. Looking closely, we'll see that it is not the stress of reproduction that kills them but rather self-destruction that is built into their life plans—a kind of planned obsolescence. Strangest of all are the animals that are able to undo their development, aging backward to an earlier stage in their life cycle, recapturing their youth, and beginning afresh in life.

When Adwaita, an Aldabra giant tortoise from India, died of liver disease in 2006, he looked every bit the same age as a young Aldabra tortoise—even though carbon dating of his shell showed him to have been born in 1750, making him over 250 years old. Some organisms undergo reverse aging, becoming stronger and more fertile as the years go on. A healthy ocean quahog, species name *Arctica islandica*, lived 507 years as measured by the bivalve's growth rings. Some clams and oysters don't seem to get any closer to death as they get older. The quahogs while alive get larger with each year of life, and annual growth rings tell a story of long lives commonly up to 400 years. Some of them were already four times as old as your grandmother when Herman Melville wrote *Moby-Dick* in 1851! Although they typically live 20 to 30 years in the wild, aging has never been detected in large-toothed fish in the class Chondrichthyes—which is to say sharks. Trees can live thousands of years, yet mayflies live one day as adults. If aging is inevitable, why such variety? What do sharks have that mayflies don't?

When to Repair Your Old Car, and When to Retire It for a New Model

It makes sense to do the small repairs but to trade the car in if repairs are major. You routinely replace the battery and the tires, but if your engine needs an overhaul, it may be more economical to buy a new car than to invest in the old one. Should we expect that Mother Nature treats living bodies in the same way—making small repairs, but starting over from an egg rather than repair major damage in an aging individual?

In the case of cars, the reason for the economic choice-point is that new cars are artificially cheap ("loss leaders," manufactured with underpaid Asian labor), while car parts are artificially expensive (they know at that point that they have no competition). Furthermore, the labor charge to take a car apart, bolt by bolt, and replace a gasket in the heart of the engine is formidable. Bodies don't have this problem, because the repair is performed on-site, cell by cell, without having to disassemble or reassemble anything.

From nature's perspective, it should *never* be cheaper to throw out an old body and start over from a sperm and an egg. The resources needed for growth of an embryo are enormous, and the failure rate (from egg to adult) is very high. The economics of a body are very different from the economics of an automobile. Why should evolution discard her tried-and-proven winners in favor of a Hail Mary pass to a tiny, defenseless newborn?

Passive Versus Active: Wearing Out Versus Self-Destruction

In fact, if the body looks as though it is wearing out with age, that is because *some* of the ways in which we age *really are passive*. The repair functions that were quite adequate in youth slow down with age, and damage is permitted to accumulate. Some aspects of aging seem to work in this way, while others look more like active self-destruction—the body is actually attacking itself.

One example of passive damage is the skin. Skin cells become damaged all the time, mostly with exposure to the sun. In youth, we have skin stem cells that generate fresh, new skin at a pace that can keep up; while in older people, there are fewer stem cells, and the stem cells we have are not working as hard.

An example of active self-destruction is inflammation, which has been found to be at the core of all the Big Three diseases of old age:

cancer, heart disease, and Alzheimer's dementia. A second mode of self-destruction is cell suicide, called apoptosis. Apoptosis can be an important function for life, as when a cell becomes infected with a virus and it falls on its sword, dissolving in a cascade of enzymes rather than allowing the virus to take over its reproductive machinery and risk infecting other cells in the body. But apoptosis is also enlisted as a program of death. In later life, some healthy and functional cells are eliminating themselves. This is what causes the wasting in our muscles (*sarcopenia*) and frailty that we suffer late in life.

I mentioned arthritis above as an example assumed to be simple wear and tear. Until a few years ago, doctors described two kinds of arthritis: rheumatoid arthritis, an autoimmune disease caused by inflammatory attack on the joints; and osteoarthritis, which is simply a loss of cartilage with age. In recent years, the line between the two has blurred.

Osteoarthritis is not wearing out but inflammation. An important defense mechanism apparently has been co-opted for the task of self-destruction. So after all, the reason our knees get creaky isn't like the reason that bearings wear out in a car.

Entropy and All That Jazz

"That's all well and good in practice. But will it work *in theory?*"

I hope I've convinced you by now that bodies are different from machines, that they don't *have to* fall apart as they get older, because they can repair themselves; that indeed everyone gets stronger, not weaker, during the time they are growing up.

But what about the theory? Isn't there a physical principle that says everything must deteriorate over time? You may be aware there is a law of nature that rules out perpetual motion machines. This is the law of *entropy*, which physicists call the Second Law of Thermodynamics. Living things are subject to the laws of physics, like all other matter. So what's up with growth and development that seem to defy the Second Law? And how does the capacity for self-repair fit into the Second Law?

Living and nonliving things both generate entropy. The way in which living things are different is that they don't accumulate entropy in their own bodies, but dump it out into the environment with their waste. Living things can take in energy, which they use to build and to repair their bodies. The Second Law of Thermodynamics applies to *sealed, isolated* systems, but living things are *open systems.* This all became clear to scientists during the nineteenth century, when the laws of thermodynamics were first formulated.

Who knew that the idea of "wearing out" could be associated with a precise, measurable physical quantity? This was the brainstorm of Rudolf Clausius (1850). Along with it came the partition of energy into the useful and the useless. The theory was formulated with the example of the heat engine always in mind, as steam was transforming transportation and industry all through Europe. Useful energy is called "free energy," and useless energy is called "entropy."

If you draw a box around a system and chart the change in entropy as different parts of the system react chemically and otherwise, you always find that entropy increases. Useful energy is degraded to useless warmth and never vice versa. Friction turns mechanical energy to waste heat. Electrical resistance is a kind of friction for electrons, with the same effect. Adding an outside source of useful energy is the one and only way that degradation (increasing entropy) can be (internally) avoided, and you can see that this is a kind of accounting trick.

This idea permitted understanding for the first time of the essential difference between living and nonliving matter. Nonliving matter is always decaying, quickly or slowly, toward a state of higher entropy. Even a steam engine must decay (wear, or rust, for example) despite the fact that energy is passing through it all the time. A steam engine has no internal mechanism to harness that energy for repair. Only living cells do that. Free energy flows in, and entropy is dumped back out into the environment. Some of the free energy is transformed into new building blocks of the cell, providing for repair and for growth.

It is tempting to think of life as having found a loophole in the Second Law, but it is really deeper than that. The capacity to repair is a feat we take for granted in living things, but it is unique to biology. No nonliving thing, no star or planet or rock or even machine, can do it. To a physicist, the capacity to use an external energy source for internal growth and repair is the very definition of life. There is no physical limit to how well the repair can be done. A rebuilt piano or truck engine can be better than new, given enough effort and attention to quality. When tiny tears appear in a muscle, or cracks in a bone, they are repaired to a standard better than new. (This is why exercise makes you stronger, and weight-bearing exercise helps prevent fractures.) A young, growing body becomes stronger, fitter, more resistant to disease, less vulnerable to mortality. So long as they have a source of free energy and a place to dump waste, there is no reason living things can't keep this trick going indefinitely.

So why don't they do it? We must look to evolution and not to physics for an answer.

For Aristotle, that venerable polymath, the best metaphors for life were a "surgeon who operates on himself" and "a ship in which the ship-building art resides in the wood." Sometimes called "the first biologist" for his close investigations of life, he was also still under the influence of his teacher Plato's notion of ideal forms. For Aristotle, life-forms have a "telos," an inherent direction or plan according to which they grow. This may be read as an intuitive account of genetics, two millennia before the study of genetics was given a scientific foundation. But during the Middle Ages, Aristotle's biology was embraced by the church, and so ideas of "teleology" in plants and animals came to be associated with an omnipotent God.

To Aristotle, death from old age was attributed to the body running out of fire (one of the four elements). He did not realize that what the body is burning is food. A fire can be kept alive as long as we are willing to throw on fresh logs. Why can't the body continue to live so long as we keep eating?

Two Misguided Theories of Aging

Since the nineteenth century, it has been known that there is no physical necessity for aging. This should have been an end to the matter, except for the confusion and basic misunderstanding that pervades the field. In the middle of the twentieth century, two theories of aging were proposed based on physics rather than biology, by people who ought to have known better. One of these became popular enough to inspire a whole subfield of aging science, including thousands of scientific papers and a theme for supplements and health food claims. This is the Free Radical Theory, and the fashionable antioxidant industry is its offshoot. The second, called Orgel's Hypothesis, made less of a splash. It was based on the fact that cells have to reproduce themselves, and every so often there must be an error in copying, and these accumulate with age. Both these theories have been discredited, though it hasn't stopped the hawking and hyping of antioxidants. Let's look at them in reverse order.

Orgel's Hypothesis, or "Atomic Aging"

British chemist Leslie Orgel (1927–2007) attached his name to this theory, but it is actually the brainchild of Hungarian-American physicist Leó Szilárd (1898–1964). Szilárd made his name studying the chain reaction that makes a nuclear bomb work, and his theory of aging was also based on a kind of chain reaction. He reasoned that as cells reproduce, they make occasional copying errors. Some of the errors may be trivial, and some severe; but each time there is an error, some information is lost, and (so he reasoned) there was no way to reverse the process or correct the error. Szilárd's theory was that copying errors multiply over time in a classic exponential curve, from one cell to two to four to eight to sixteen . . . This is like the slow start and rapid rise in the chain reaction of an atomic explosion: "atomic aging."

When a physicist looks at aging as an abstract process, the striking feature is that species have a set life span, and deaths tend to cluster around that age. By contrast, when manufactured objects wear out, they tend to fail more randomly, spread over a longer period of time, with some fraction surviving many times longer than the average. For example, the average useful life of a car in America is twelve years, but there are many cars on the road over twenty years old and a handful that are thirty years old and still functional. The current life expectancy of Americans has grown to seventy-eight years, but there are zero Americans who are twice as old as that. An attraction of Szilárd's theory was that it could account for the fact that aging takes a small toll in young people, but accelerates over time and becomes a wall of death.

The Szilárd/Orgel theory lost a lot of its appeal once stem cells were discovered. The relationship to stem cells may not be obvious, but it makes for a good story because the reasoning is clear, and scientists did the right thing—abandoning a theory when it no longer made sense.

In the 1960s, when Szilárd and Orgel were working out their theory, it was assumed that the body's tissues grew cell by cell. It was logical to imagine that skin cells made new skin cells and muscle cells made new muscle cells and liver cells made new liver cells. On this basis, they assumed that any copying errors would be spread more and more widely in each generation of cells.

Stem cells were first discovered in 1978. It gradually became clear that new muscle cells don't come from old muscle cells or new skin cells from old skin cells—rather, they all come from stem cells. Just as there are cells that specialize in functions like skin and muscle and liver, stem cells are cells that specialize in reproduction. They are the queen bees of the body, and their progeny can grow up to be whatever they want to be. You may hear the word *pluripotent* in descriptions of stem cells. It means that stem cells have the potential to generate progeny of many different types.

The system of using stem cells offers the body a solution to the problem of error accumulation. It means that there are no copies of copies of copies with ever-increasing numbers of errors in each generation. Instead, the body keeps stem cells as pristine originals, so errors don't accumulate. It is as if evolution anticipated the problem that Szilárd identified and built in a system to avoid it, probably hundreds of millions of years ago.

The Orgel Hypothesis died a graceful death, and its tomb was sealed in 1980 when a young scientist named Cal Harley actually counted the copying errors from one cell generation to the next using early DNA sequencing technology. He found that there was no appreciable accumulation. Biological replication can be amazingly accurate when it wants to be. Human DNA is copied from one generation to the next with about one copying error out of ten billion units.

Leó Szilárd (1898–1964) was one of the most brilliant and important scientists you have never heard of. The Hungarian-American physicist first suggested the possibility that nuclear fission could be realized as a self-sustaining chain reaction. In other words, it was Szilárd who proposed that a uranium atom that split into two smaller atoms could be a trigger, nudging other uranium atoms to do the same thing, and the result could spread quickly through a hunk of uranium, causing a substantial release of energy. As a Jewish refugee from Hitler's Europe, grateful to America, Szilárd played a key role in the Manhattan Project during World War II; but he personally urged President Truman not to use the superweapon he had helped create against the Japanese people but to demonstrate it instead in Tokyo Harbor. After the war, Szilárd helped found the Council for a Livable World, an early, high-profile disarmament advocacy group, and he spent the last years of his life studying biochemistry, not physics.

The Free Radical Theory of Aging

This is another theory of aging that came from outside the field of evolution, also based on abstract thinking and physics but minimal biology. Its greatest claim to success has been in birthing the market for antioxidant supplements. The year 1956 was a time of atom bomb tests, radioactive fallout, and a very real public fear of nuclear war. Working in a UC–Berkeley medical lab, a young theoretical chemist named Denham Harman studied the effects of radiation on mice and noticed that it made young mice appear old. He cataloged some kinds of damage that radiation could do to sensitive molecules, and it occurred to him that high-energy chemistry inside the cell might do the same kind of damage. His experiments showed that antioxidants could protect mice from some of the damage of radiation, and he proposed that perhaps antioxidants could slow the aging process.

This is a theory that medical gerontologists still apply today, though, as we shall see, it has faced major setbacks and lost a lot of its credibility. It is a theory about the "how" of aging, rather than the "why," and this should have been a cause for suspicion from the start. It describes one kind of damage that the body suffers but tells us nothing about why the body should be failing to repair itself or why species should have evolved to permit this damage to accumulate.

The Free Radical Theory gave birth to the antioxidant craze, and the embarrassing failures of antioxidants to work in clinical trials have undermined support for the Free Radical Theory, though they have done little to quiet advertising claims for antioxidants as a path to longevity.

In every cell of our bodies, the chemical energy of sugar is processed for the cell's use in hundreds of tiny islands called *mitochondria* scattered through the cell. Not surprisingly, the mitochondria use molecules that carry a wallop in order to store energy efficiently. There are by-products of these high-energy reactions, half-formed pieces of molecules called *radicals* that aggressively look for some other molecule to combine with.

"Free" radicals are simply radicals that have not yet found anything to latch onto. Free radicals behave like loose cannons in the ordered, civilized world of the cell's biochemistry. They attack the complex, delicate, and perfectly formed molecules that do the work of the living cell and convert them to useless, damaged forms.

Harman's theory of the free radical argued that damaged biomolecules accumulate over time, causing aging. In favor of the theory, cells deploy an army of chemical traps to quench the free radicals before they can do damage, including glutathione (GSH), superoxide dismutase (SOD), and ubiquinone (CoQ10). However, damaged biomolecules do accumulate in aging cells. And mitochondria become less efficient with age. If we feel less energetic than we used to, it's because the energy factories in our cells are compromised.

Denham Harman, the industrial chemist who posed the Free Radical Theory in 1956, was unable to increase maximum life span with antioxidants, ultimately realizing that they do not enter the mitochondria, the main source of free radicals. In fact, recent evidence shows the exact opposite of the proposed theory—to wit, that oxidative stress promotes longevity. It's a theory even more radical than the Free Radical Theory; abundant and accumulating evidence suggests that aging is a genetic and physiological "inside job." More than free radicals are involved in senescence: apoptosis (programmed cell death) is implicated in dementia (which slowly destroys our brains) and sarcopenia (which slowly destroys our muscles); "purposefully" limited telomerase, which the body can easily make but doesn't always, is implicated in cell senescence—the limited number of cell divisions of our body cells. Evolution has found that killing its own, letting or making it happen, can be useful. But how and why would evolution do this to organisms with a survival imperative? The answer only makes sense once we step out of the foundering neo-Darwinian paradigm.

A growing minority of evolutionists is broadening the neo-Darwinian framework to include symbiosis, entangled social networks, and the emergence of new levels of organization in evolution.

Free radicals are powerful oxidizers. The damage that they do is *oxidation*. The protective chemical traps that neutralize free radicals are antioxidants. In the 1970s, in response to the Free Radical Theory, there was excitement among researchers about the potential of antioxidants to enhance health and extend life.

But failures piled up from the very beginning. It was true that more oxidative damage could be seen in older animals than younger animals, but antioxidants did not seem to protect the cells or make lab animals live longer. The theory continued to attract support because the idea was so appealing, despite the lack of experimental corroboration.

What does oxidation mean? All molecules are made of atoms held together by chemical bonds. There are two kinds of bonds. Low-energy covalent bonds occur between similar atoms that share a pair of electrons. High-energy ionic bonds occur between very different atoms, where one steals the electron from the other and becomes negatively charged, while the first becomes positively charged, and thus they stick together. Electron sharing is a low-energy bond. Electron transfer (or theft) is a high-energy bond. Oxygen is an electron thief that can rearrange molecules with low-energy covalent bonds, substituting high-energy ionic bonds; and when this happens, a lot of energy is released (oxidation). Living cells tend to use ionic bonds for energy metabolism and covalent bonds for everything else.

Despite failures in animals, antioxidant vitamins were considered safe and promising enough for human trials. In the 1980s and '90s, the Alpha-Tocopherol, Beta-Carotene Cancer Prevention (ATBC) Study was a huge experiment based in Finland, with tens of thousands of male smokers assigned randomly to receive antioxidant vitamins or placebo pills. In 1994, the experimenters first reported that the cancer rates and the death rates were both *higher* in the men who had received antioxidants than those who had received the placebo. Researchers were completely puzzled and had no explanation and suspected that the result was an unidentified methodological error, because it was too big to be a statistical fluke. In 1996, the experiment was halted, because the investigators realized that antioxidants were killing their subjects, though they still did not understand why.

With two narrow exceptions, all the dozens of studies using antioxidants to try to extend life span have failed. The exceptions used glutathione and a Russian-designed molecule nicknamed SkQ, both targeted to the mitochondria.

An early gene discovered in worms was named CLK-1 because, when the gene was deleted, the worms lived 40 percent longer. CLK-1 means "clock 1." Over the ensuing years, other animals were discovered to have their own versions of CLK-1. Fruit flies and yeast cells and even lab mice were found to live longer when they were deprived of the gene homologous to CLK-1. This gene is needed to make coenzyme Q ubiquinone, or (CoQ). This is the first-line antioxidant of the mitochondria themselves. CoQ has been promoted for decades as an anti-aging supplement (and yet it might have some value for heart patients). But it's a surprising fact that animals genetically deficient in CoQ tend to live longer.

Why Did Antioxidants Fail So Spectacularly?

Nietzsche said, "God is dead, but the news has taken a long time to reach man." The same may be said of antioxidant studies. Twenty years after these failures, the question of why is still being unpacked, but one basic and unexpected twist in the theory seems to be central. Free radicals, these very same free radicals that are the source of damage, are used by the metabolism as intercellular messengers, sentinels that call for a state of high alert and put the body's defenses into overdrive, with the result that we live longer.

Exercise generates copious free radicals, and yet exercise makes us live longer. In fact, there is a line of research suggesting that antioxidants can blunt the benefits of exercise.

But surprise—it turns out that free radicals, these savages in the civilized world of the cell's chemistry, have been recruited as couriers, and one message they carry is, "Repair the infrastructure for a long and healthy life!"

Instant Replay

It is seductive to think that the body wears out the way a machine wears out. But there are key differences between living and nonliving matter. Living things are not destined to the same sort of decay that governs machines, because living things have an energy source and a capacity for repair. Life as a whole has endured and expanded for some four billion years without wearing out. There is no physical reason why an old body can't be stronger and fitter than a young one. And in reality, we will see that there are some animals (and many plants) that don't age at all but keep growing larger and stronger indefinitely. Our bodies clearly are not doing their best to keep themselves in good repair or to live as long as possible. We know this because life expectancy is short when the damage is least and the energy available to repair the damage is greatest. In addition to this neglected maintenance, there are some ways the body *actively* destroys itself later in life. The bottom line is that we can't think of aging simply as a mechanical or chemical process of wearing out. And the first application of this insight is that all those antioxidants that we have been sold to protect our delicate chemistry from damage might be doing more harm than good.

TWO

The Way of Some *Flesh:* *The Varieties of Aging Experience*

The handsome young man, the lovely young woman in their prime—Death comes and drags them away. Though no one has seen Death's face, or heard Death's voice, suddenly, savagely, Death destroys us, all of us, old or young. And yet we build houses, make contracts; brothers divide their inheritance, conflicts occur, as though this human life lasted forever. The river rises, flows over its banks, and carries us all away, like mayflies floating downstream. They stare at the sun then, all at once, there is nothing.

—*Gilgamesh,* from Mesopotamia,
one thousand years before the Bible

Humans age gradually, but some animals do all their aging in a rush at the end of life, while others don't age at all, and a few can even age backward. The variety of aging patterns in nature should be a caution sign to anyone inclined to generalize. The fact that aging doesn't occur in some organisms suggests that it can be stopped in others that currently undergo it.

Bacteria reproduce symmetrically, just dividing in two. What could "aging" mean for bacteria since, after reproduction, there is no distinction between parent and child? Single-celled protists (like the amoeba) also reproduce symmetrically, but curiously, they invented a way to age

nevertheless. It is called "cellular senescence," and we will have much to say about it in chapter 5. And even among macroscopic life-forms, life spans of organisms are immensely variable in a way that is finely tuned to local ecologies and reproduction rates. This can hardly be the result of a universal, inexorable process; in fact, such fine-tuning to circumstance is the signature of an adaptation.

Two Kinds of Microbes: Bacteria and Protists

Protist is the formal name for complex one-celled life—cells with nuclei ("eukaryotic cells") like plants or animals but not plants or animals. Taken as a whole, all protist species (e.g., amoebas and ciliates) are called protoctists, which also includes multicell versions. (Some two hundred thousand species of protoctists are known.) Bacteria are also one-celled life-forms (although they can form multicell alliances), but they are simpler and much older. Protists, or protozoa, are about a million times bigger and much more highly structured than bacteria, with a cell nucleus containing DNA and many separate functional parts, called *organelles*, all within a single cell.

Bacterial life probably arose about four billion years ago (unless it came from space, in which case, it may be much older). Protists are only about two billion years old, while plants, animals, and fungi are only about half a billion years old.

Bacteria have little internal structure in their cells. Bacteria don't have true chromosomes. Rather, their DNA comes in rings called *plasmids*, which are spread all through the cell. Protists have a great deal of cell structure. There are communications and transportation networks (*endoplasmic reticula*) spread through the cell. Just as your body has liver and kidneys and stomach and heart, protists have organelles that perform different functions, from energy production in the mitochondria to waste recycling in the lysosomes.

Most of protist DNA resides in the cell nucleus; a little bit in the mitochondria—or, if they are algae, in the organelles where

photosynthesis takes place, called *plastids*. Some protists have two-cell nuclei. The micronucleus contains one archival copy of the DNA, while the macronucleus contains many working copies that are continually being transcribed in order to drive and to regulate the cell metabolism. Although you may never have heard the word, protoctists are of great importance in evolutionary theory because our kind of multicellularity, sex, and aging evolved in them.

Life spans range from Methuselans great and small to genetic kamikazes that die of a spring afternoon. Why do some organisms die right after they reproduce while others' lives unfold at a stately pace? Mayflies spend many months as nymphs, growing slowly underwater, but as soon as they reach maturity, they mate and die within hours. Aspen trees can propagate underground for thousands of years. Yeast cells live for days, bowhead whales for centuries, but some of the genes that control aging are common to yeast and whales.

And it is not only the length of life but the pattern of deterioration within that time that varies widely. Aging can occur at a steady pace through the course of an entire lifetime (most lizards and birds), or there can be no aging at all for decades at a time, followed by sudden death (cicadas, sea birds, and century plants).

Genetically directed death is linked to reproduction in many but not all species. Pansies are colorful varieties of the species *Viola tricolor*. Normally, they die after flowering. Some would say they've shot their wad, put their last ounce of energy into reproduction, and they die from a kind of exhaustion. But gardeners know that if they snip the flowers, then new flowers will grow in their place. Snip again, and they grow back again. We can keep the pansies flowering all summer, but the week that we let the pods go to seed, the pansies shrivel and die. This belies the "exhaustion" theory. The plants have plenty of energy (and genetic know-how) to produce generation after generation of flowers! Their death seems to be triggered by the formation of the seed pod, a matter of signaling rather than exhaustion. This suggests that there was never

any necessity for the plants' death, that Mother Nature (a.k.a. "natural selection") had a choice in the matter and chose death over life.

Some organisms have an indefinite life span. They just go on living until something kills them, but they don't deteriorate or become more likely to die over time. Blanding's turtles and *Sanicula* bushes are like this. Stranger yet are animals and plants that are *less* likely to die over time. Sharks, clams, lobsters, and most trees just keep getting bigger and stronger with each passing year. But maybe this isn't so strange—if aging weren't so familiar a fact of life, perhaps we would think it perfectly normal that creatures should be able to marshal their resources and consolidate their strength over time. Biologist Stephen Jay Gould once remarked that if we were octopuses, we would worship a God with eight arms. Such octopuses would also not believe in gradual, inevitable aging, since octopuses (as we'll see on page 71) are unable to feed after they lay their eggs.

Our own "inner assassin" works with stealth, like an evil empress gradually poisoning her husband; but other species have inner killers that do their deed far more quickly, and still others appear to have no genetic death programs at all. Such variety is a sure signal for a feature molded by active natural selection, not an immutable law of entropy.

The Hyperboreans, a mythical Greek race who live beyond (hyper) the north wind (Boreas), never get old or sick but frolic about all day with golden bay leaves in their hair. Though they don't die of old age, they are not immortals, and chance conspires to kill them, one by one in due time.

This is non-aging, and it sounds pretty strange. But a real-life example is found in some plants. *Sanicula* is a shrub growing in the meadows of Sweden, and one plot in particular has been studied continuously for sixty-five years. *Sanicula* does not age. About one shrub in seventy dies each year, so that the plants have an average life span of seventy years. However, as death is merely a matter of constant chance, a seventy-year-old plant has no more mortality

risk than a ten-year-old plant. With one plant in seventy dying each year, there will be about half left at the end of 50 years; but the plants in that half will be as fresh and young as they ever were. At the end of another 50 years, a quarter still remain, an eighth are alive after 150 years. At this rate, about one Swedish shrub in a million should live a thousand years.

"Aging"—A Demographer's Definition

What does it mean to say that members of a certain species age rapidly, or slowly, or not at all? What do we mean when we say that "he looks young for his age" or that "she aged ten years when she went through the divorce"?

As the biomarkers of aging vary widely from one species to the next—indeed, from one *individual* to the next—it's difficult to come up with a single universal definition. A man may be prematurely gray, and a naked mole rat baby may be covered with wrinkles. For the actuary, however, the question has a clear answer, even if it's one only a statistician could love: aging is an increase in the mortality rate. In other words, as an animal gets older, it suffers an ever-higher risk of death.

For example, a twenty-year-old man has a 99.9 percent chance of living to see his twenty-first birthday. This is to say that his chance of dying is one in one thousand per year. If this were to continue, then a forty-year-old would also have a one in one thousand chance of dying before his forty-first birthday. We'd call that "no aging." In reality, a forty-year-old has a two in one thousand chance of dying before his forty-first birthday. This doubling of his mortality risk over twenty years is evidence of gradual aging.

It gets worse. The risk for a sixty-year-old is ten in one thousand and for an eighty-year-old, sixty in one thousand. (These figures all come from the Social Security actuarial tables for 2010.)

Age (male)	Prob of death
20 yrs	0.001
40	0.002
60	0.01
80	0.06
100	0.36

Mortality Table (Your Chance of Dying)

Not only does the risk of dying increase, it increases faster and faster. This is called "accelerating senescence." But other species have different patterns. The probability of death may increase and then level out: "decelerating senescence," or it may even decrease. If the probability of death doesn't increase, then the species doesn't age at all. It is consistent, if stranger, to say that if the probability of death goes down from one year to the next, then a species is aging backward, which is "negative senescence."

Another Definition of "Aging"

There's a second objective measure of aging, and that is decline in fertility. Just as mortality is defined as the probability of death, fertility is defined as the probability of reproduction. Men lose fertility gradually over their adult life. Women lose their fertility more rapidly, and fertility drops to zero at menopause. But different species have different patterns, different schedules. In some species, fertility *increases* over much of the life span, another form of "negative senescence." For example, Blanding's turtle, a species of box turtle common in the American Midwest, matures slowly over decades, and it doesn't keep growing, but it does continue to increase in fertility. Its risk of dying also declines with age.

From an evolutionary perspective, the loss of fertility is primary. From the perspective of natural selection, once you're no longer able to reproduce, you might as well be dead.

Scientists Are Human—Must Aging Exist? Hamilton's Proof and Vaupel's Spoof

Caltech physicist Richard Feynman warned his students, "The first principle is that you must not fool yourself—and you are the easiest person to fool." Ironically, it is the greatest minds at the height of their success whose lapses travel farthest and do the most damage. That's because these people have earned a level of respect that makes other scientists shy about reviewing their manuscripts critically or denying publication.

So it is no reflection on the greatness of William D. Hamilton, either as a scientist or as a person, that he fell foursquare into a generalization about aging that we can only regard as risible. Hamilton proved something that is objectively not true—that all organisms age. But such is Hamilton's reputation (based on the many things he got right), that his erroneous "proof" that aging *must* exist continues to be cited as gospel almost fifty years after the fact.

In 1966, Hamilton published a proof that the kind of gradual aging experienced by humans is a requirement of evolutionary law, based on some general and plausible assumptions about the way natural selection works. Young Hamilton was smart and courageous enough to make bold predictions based on mathematical logic. To his further credit, he was honest and confident enough to change his perspective late in his life, when evidence warranted.

William D. Hamilton (1936–2000) was a wild character, acclaimed for his mathematical formulation of kin selection, and politely ignored when he scoured chimp feces in Africa for evidence of an origin of the AIDS virus in the oral polio vaccine. Egyptian-born world traveler and intrepid naturalist, he suggested that clouds

might be distribution systems for microbes. In his later life, he became fascinated by the microbial protection that sexual diversity provided to populations and championed the Gaia Hypothesis—that organisms acting together can physiologically regulate their environment. Long after the equation known as "Hamilton's Rule" became the canonical formulation of the Selfish Gene paradigm, he came to appreciate that cooperation and, yes, group selection has played an essential role in evolutionary history. The second volume of his only published book, *Narrow Roads of Gene Land*, is called *Evolution of Sex* and is devoted to the Red Queen hypothesis, explaining how natural selection insisted upon the sexual blending of genomes despite its manifest disadvantage for individual selfish genes. That book follows his intellectual journey, from a focus on the mathematics of individuals to a closer look at how groups can work together to survive. After suffering for decades from malaria and *Giardia* that he picked up in his field work, Hamilton died with his boots on, suffering a hemorrhage after his last trip to Africa.

Right at the beginning of his scientific paper, Hamilton claimed "that senescence is an inevitable outcome of evolution," that "cannot be avoided by any conceivable organism." Hamilton's "proof" was derived within the context of the prevailing theory of his day. The origin of aging was thought to be in genes that act at one age only. (This was a common assumption before there was a science of gene expression, or "epigenetics.") His proof compared the effects of natural selection acting at different stages of life. Selection is stronger on the genes that activate earlier. What Hamilton showed was that as long as you're going to die, it's better to die later than sooner, so evolution has concerned itself more attentively with things that kill us when we are young. (This insight was Peter Medawar's starting point when he launched the modern theory of aging in 1952. More about this in chapter 4.)

Hamilton was thinking mathematically while avoiding the elephant in the room. The particular elephant he missed was the effect of size.

Hamilton was habituated to thinking of humans and dogs and cats (and ants and elephants and most land animals), which grow to a fixed size at maturity and then stop growing. We might forgive him for neglecting sea urchins and clams and lobsters and (maybe) sharks, but why wasn't he able to see the trees? Like trees, some animals have no fixed size.* They keep growing larger all through their lives. With larger size comes greater capacity to make seeds, so their fertility goes up. With larger size comes strength to resist the wind and the weather (in trees), and to deter predators (otters love to eat sea urchins, but big ones are harder to sink your teeth into); so death rate decreases with age. And that's backward aging according to the standard definition, the same one Hamilton used to conclude that backward aging is impossible.

In his densely argued thirty-four-page paper full of equations and syllogisms, William D. Hamilton concluded that it was inconceivable for any known organism to avoid aging. However, living organisms continue blithely to thumb their noses at Hamilton's theory (if a tree may be said to thumb its nose).

James Vaupel, director of the Max Planck Institute for Demographic Research in Rostock, Germany, may not be a household name, but he is the foremost demographer of our time. He is best known for documenting the steady rise in human life span in the developed world, which has proceeded at a remarkably even pace for over 160 years. Perhaps it was just to make a point that Vaupel worked with his student Annette Baudisch to create their own proof (spoof?) in counterpoint to Hamilton's. In a provocative 2004 article, they offer a general "sproof" that *aging is impossible.* From the same postulates that Hamilton had used forty years earlier, Vaupel and Baudisch "sproved" that the probability of death must always *decrease* with age. Vaupel reminds us that population biology is an experimental science, where theory is always tentative and must be checked against reality.

The assumptions that go into Vaupel and Baudisch's proof are not only very reasonable, they closely parallel the reasoning in Hamilton's proof. What Vaupel and Baudisch prove is that it is always worthwhile

* There are physical reasons that animals can't grow too large. Carrying around a lot of weight becomes inefficient. Just look at the relative proportions of an elephant's legs compared to the legs of a gazelle. This isn't such a problem for animals that live in the water, as the example of great blue whales suggests.

for the animal to invest in the future. It never stops being true that there is a benefit to building stronger muscles, greater size, a more robust immune system, and so on. ("Growth" is generalized to stand in for all these things. It is an investment in greater resistance to environmental threats, for now and for the future.)

Hamilton's proof shows that aging must always evolve. Vaupel and Baudisch's proof shows that it is impossible for aging to evolve. Hamilton proved fertility must always decline and mortality increase with age. Vaupel and Baudisch proved that fertility must always increase and mortality decline with age. I suspect that this delicious irony was exactly the goal that they had set in the sights of their Big Science gun.

Vaupel is, by background and inclination, a demographer. He collects and analyzes statistics about population and mortality. Unlike most evolutionists, he cannot be intimidated by math. But he is not a theorist in the tradition of Hamilton, seeking to create grand, unifying frameworks for understanding evolution. So Vaupel is content to sow doubts amid Hamilton's assurance and leave us with the paradox.

If we are theoretically inclined, we might try to examine the wreckage and ask what went wrong in both these proofs—because clearly both proofs are fallacious. There are, in nature, both creatures that grow weaker with age and creatures that grow stronger with age, and all these creatures were molded in a process of natural selection. Why has natural selection included aging in so many life plans, despite the fact that aging is anti-fitness? Why do we find aging and non-aging and even reverse aging in nature?

Aging Trajectories

Humans age on an accelerating schedule. By the late teen years, some aspects of our fitness edge are already eroding. But changes are much more rapid in the seventh decade, and thereafter, the losses cascade and the vulnerabilities pile up. This is just one kind of aging schedule, and not even the most common one in the animal kingdom. Aging can be gradual or sudden. Vulnerability to death can increase or decrease over time, or it may remain constant ("negligible senescence"). Aging can level out, *decelerating* instead of accelerating. Albatrosses and naked mole rats remain completely healthy through their lives and then die at a predetermined time. Finally, there are animals capable of reverting from their adult stage back to larval form, from whence they came. When their bodies sense that the environment is so hostile that it was probably a mistake to grow up, they reverse the process and become larvae once more. This can only be considered "rejuvenation" or "aging backward."

So the shape of the "aging trajectory" can have every possible pattern. Moreover, the time scale on which aging unfolds can also be as short as a few hours or as long as hundreds of years (for long-lived animals) or thousands for long-lived plants. The very first thing that any theory of aging must account for is the huge variety of time scales and contours of the aging curve in nature.

New England lobsters today are a restaurant delicacy, priced higher than a sirloin steak. They are airlifted live to Japan, where they command an even higher price. But in the nineteenth century, lobsters were so copious in New England that they were discarded as bycatch. The state of Massachusetts dealt with the excess by supplying lobsters as daily prison fare, until prisoners went on strike and refused to eat them. New England lobsters today are fished so heavily that they rarely grow larger than a pound, but lobsters weighing more than ten pounds are still caught occasionally (and usually released). The largest lobster on record was forty-four pounds. The reason that the large lobsters are released back into the ocean is not just that they won't fit on a dinner plate. Lobsters become more fertile as they grow larger, and their young are more viable. A few large lobsters can be the breeding stock for a large area. We don't have an age record for the oldest lobster ever caught because lobsters don't have annual rings or layers that broadcast their age. The forty-four-pounder was said to be more than one hundred years old, but no one knows for sure.

Clams also can grow larger and more fertile indefinitely. But clams have growth rings that count the years for us. The oldest clam on record (an ocean quahog of the species *Arctica islandica*) has been tagged at 507 years. Small clams have natural predators, including starfish that latch onto their shells and pull them apart by brute force. But once a clam outgrows the arms of a starfish, it can keep growing indefinitely. Clams have one foot, one mouth, no eyes or ears or stomach, no brain. Giant clams, up to 750 pounds, live the same lifestyle as their smaller relatives, sucking in the seawater, taking in thirty thousand times their weight in water every day, and filtering out plankton and algae, which continue to grow and reproduce inside them. Such clams can be golden brown, yellow, or green, gaining nourishment from the photosynthetic beings lodged within them. Like giant lobsters, the giant clams provide eggs for a whole community. They have been known to release half a billion eggs in a day.

The rougheye rockfish grows in deep, cold water off the North American West Coast, from San Diego up through the Aleutians, and down the Asian coast to Japan. At ocean's bottom, light is absent, food and oxygen are scarce, and temperatures are low. (Because of high pressure and salinity, the temperature can actually be below zero Celsius, which is normally the freezing point of water.) Everything seems to happen more slowly there and, perhaps appropriately, the rockfish life cycle unfolds at a leisurely pace, with some living more than two hundred years. Like other victims of overfishing, rockfish are now an endangered species. Since their life cycle is so long, it would take decades for them to recover under the best of circumstances, but given the general devastation of our oceans, they don't stand a chance. Apart from their intrinsic value as long-evolved beings, their extinction may deprive us of clues to the secrets of aging.

The Procrustean Perspective of Annette Baudisch

We find it natural to classify different species as living a long or a short time, to lump together the insects that live for a day and distinguish them from the trees and whales that live hundreds of years. But much of that difference can be attributed to size. Everything from growth to reproduction to aging must occur more slowly in a behemoth with a slow metabolism and tons of tissue to nourish. So we are inclined to be more impressed with a honeybee that lives twenty years than we are with a moose that lives twenty years.

But suppose we were to remove length of life completely from consideration and compare different species based on the *shape* rather than the *duration* of their life histories. Rather than asking how long they live, ask instead whether their populations tend to die out gradually, or if many die in infancy and fewer later on, or if all the deaths bunch up at the end of the life cycle. This indeed was the brainchild of Annette Baudisch, a student of James Vaupel whom we met in connection with the sproof that aging cannot exist. Vaupel's brilliant student went on to introduce a new lens through which to look at the comparative biology of aging.

This procedure opens a new window, a new way of looking at aging across species. A chart published in a paper in *Nature* in 2014 applies the methodology that Baudisch had pioneered a few years earlier. What emerges from this picture is the breadth of nature's ingenuity. Every conceivable combination is represented, with rapid aging and no aging and backward aging, paired with life spans of weeks or years or centuries. The strange bedfellows that appear as neighbors on the chart are utterly unexpected. For example, at the top of the chart, with low mortality that rises suddenly at the end of the life span, humans are joined by lab worms and tropical fish (guppies)! In fact, in terms of aging profiles, we humans look more like the lab worm than the chimpanzee.

One Size Fits All

In Greek mythology, Procrustes was a bandit who invited travelers into his guest room, assuring them that he had a bed that was just the right size for them. If they were too tall, he cut off their feet, and if too short, he stretched them on a rack.

The graphs on the next page show the varieties of ways that animals and plants age in the wild. The light downward line in each frame is the survival curve, and the bold curve underneath is fertility. The downward slope of the survival line just means that fewer and fewer individuals are left alive as time goes on. The way this graph has been constructed, a straight line going diagonally downward [\] is neutral, or no aging at all—for example, the hydra and the hermit crab. (The hydra is like a freshwater jellyfish, a quarter-inch long and found in ponds.) Lines that hump above the diagonal [ ] represent normal aging, while lines that dip below the diagonal [ ] represent reverse aging, or "negative senescence." All the animals in the top row show "true aging"—they are more likely to die as they get older. The next two rows show plants and animals that don't age or that age in reverse. For the latter case, the older they are, the less the risk of death. Most trees are like this, and tortoises follow the same pattern, as do clams and sharks (not pictured). But for animals and plants in the bottom two rows, the death rates are steadier. For turtles and oak trees, in fact, the curves flatten out. That means that there are fewer of them dying old than dying young, which is aging in reverse.

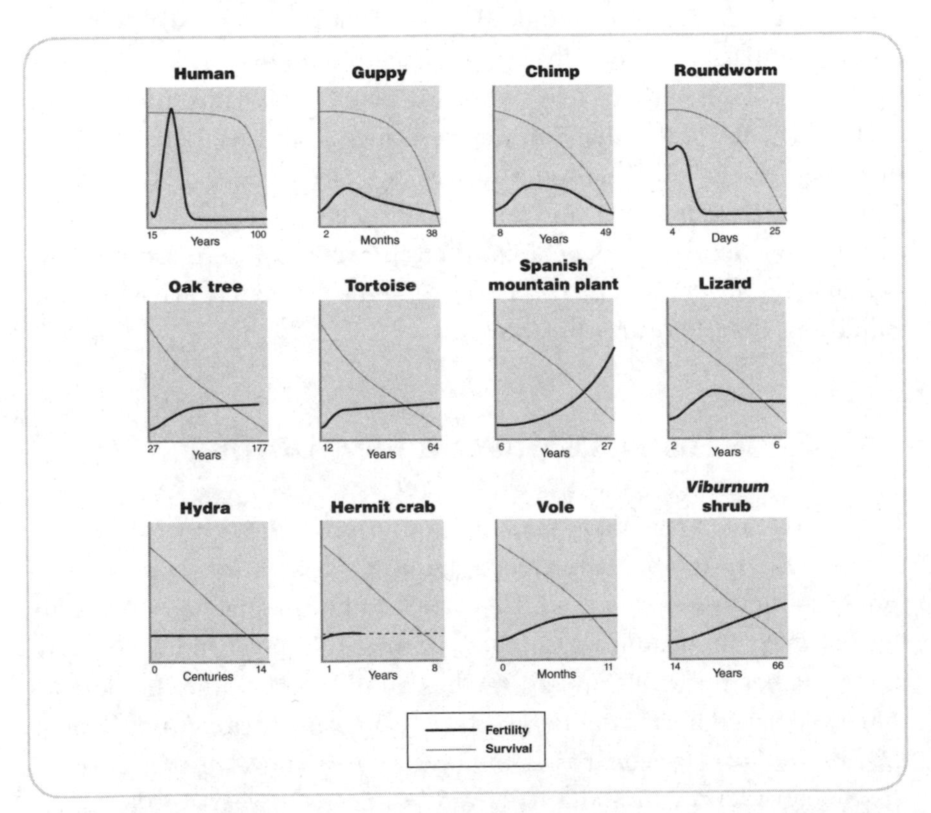

Fig. 1. Aging: A Procrustean Comparison. Aging and fertility plotted for a range of organisms. There is no monolithic aging or fall-off in fertility among all of Earth's multifarious beings. The dark line, representing fertility, sometimes rises with age. The light gray line, survival, represents decreasing chances of living with each year—but we can see that sometimes (e.g. for the tortoise and the end of the lizard's life) it does not proceed straight down but turns up, showing an increase in chances of survival relative to previous years. (Adapted from Jones et al, 2014.)

The bold line, representing fertility, is straightforward. Fertility may rise as an animal or plant grows bigger, or it may fall with reproductive aging—for example, in menopause. Notice that animals in the top row lose all their fertility well before they are dead. This poses an evolutionary conundrum in its own right. If the sole target of natural selection is to maximize reproduction, then why has evolution allowed reproduction to fall to zero while so many remain alive? The rising fertility curves indicate increased reproduction with age, which is another kind of negative senescence. When you think about a tree that grows larger with each passing year, it's not so surprising that it's making more seeds the older it gets. The Spanish mountain plant in the second row is *Borderea pyrenaica,* a plant that grows out on the rocky cliffs of the Pyrenees mountains. If undisturbed, it can live to three hundred years or more with no sign of aging; but notice that its fertility doesn't really get going until it is more than twenty years old; perhaps that's why this plant is critically endangered.

The message of this diagram was described in the accompanying article: nature can do whatever she wishes with aging (or non-aging). Any time scale is possible, and any shape is possible, and each species is exquisitely adapted to its ecological circumstance. There are no constraints. Yet all the accepted theories of aging today are based on assumptions that there are constraints.

Instant Aging, Sudden Death

The antihero of Oscar Wilde's famous gothic novel *The Picture of Dorian Gray* is able to temporarily escape aging, because a magic portrait of him is doing the aging instead. He parties and womanizes and ruins the lives of everyone whom he touches, until at the novel's end, his hard living catches up with him and he dies an old man, while the portrait that has delayed his mortality reverts to its pristine form. Frank Capra's 1937 movie *Lost Horizon,* based on James Hilton's novel, is the source of the Shangri-La myth. Maria has lived hundreds of years in the Valley and retains her youth and beauty, but when her British paramour convinces her to depart from the Valley, she turns into an old hag, loses all her strength, and dies within hours.

Apparently, life imitates art. In nature, aging to death can be rapid and sudden at the end of a reproductive cycle. Sudden post-reproductive death is common in nature, affecting organisms as varied as mayflies, octopuses, and salmon, not to mention thousands of annual flowering plants. Biologists refer to this life story as "semelparity." Though the etymology is from the Latin for "single birth," the word reminds us of a Greek myth.

Semele, a mortal woman, had an affair with Zeus, inciting the ire of his wife, Hera, who, perhaps playing to his cockiness, extracted a promise from the Olympian king that he would reveal himself to his paramour in his full glory. The result was a lightning-storm bout of ardor for which the body of poor Semele was not prepared; indeed, it led to her death, although not that of her unborn son, Dionysus, god of wine and orgies, whose fetus Zeus recovers from Semele's charred womb and sews into his thigh; hence Dionysus's appellation as the "twice born."

The cause of death in semelparous organisms varies widely. Octopuses just stop eating. Praying mantis males make an ultimate reproductive sacrifice, giving themselves up as snacks to their female partners. Salmon destroy their own bodies with a blaze of steroids. These modes of death are so obviously genetically programmed that even some of the most orthodox neo-Darwinists are compelled to regard them as an exception to the thesis that programmed death is an evolutionary impossibility.

Some opossums and fish undergo multiple simultaneous organ system failure after reproducing. Chinook salmon hatch in river pools, often hundreds of miles upstream from the sea. They spend their first year or two in the protected environment of the river, where life is tamer and large predators rarer. When they have grown large enough to compete, they migrate downriver, out to the ocean to seek their fortunes. They may range up to 2,500 miles from the mouth of the stream where they first entered the sea. They live in the ocean anywhere from two to seven years, growing larger but not weakening or becoming frail with age. When they are ready to reproduce, they find their way back, not to any handy river mouth but to the very same river pool where they were hatched. Their journey is a headlong rush, simultaneously into fertility and death.

By the time the adult salmon reach their spawning ground, their metabolisms are in terminal collapse. Their adrenal glands are pumping out steroids (*glucocorticoids*) that cause accelerated—almost instant—aging. They've stopped eating. Moreover, the steroids have caused their immune systems to collapse, so their bodies are covered with fungal infections. Kidneys atrophy, while the adjacent cells (called interregnal cells, associated with the steroids) become greatly enlarged. The circulatory systems of the rapidly deteriorating fish are also affected. Their arteries develop lesions that, interestingly, appear akin to those responsible for heart disease in aging humans. The swim upstream is arduous, but it is not the mechanical beating that fatally damages their bodies. It is rather a cascade of nasty biochemical changes, genetically timed to follow on the heels of spawning. The symptoms affect both males and females, despite the uneven share of metabolic work that falls to females, whose eggs may constitute a third of their body mass during the final leg of their trip.

Sex for salmon involves a meticulous choice of mate but no actual touching. It is more like "parallel play." The female chooses a male partner and then digs a ditch ("redd") in the gravel stream bottom, turning slightly to one side and using her tail for a shovel. Side by side, the couple deposits eggs and sperm in the redd. Salmon are not so imprudent as to put all their eggs in one basket. After their first mating episode, they move together through the pool, choosing other locations, digging more redds, repeating their dance and their parallel play until they've shot their respective wads. The female carries eggs that are far larger than the male's sperm cells, and she is exhausted sooner than her mate, so the male may indulge in polygamy and seek another mate. Eventually, male and female have exhausted their stocks, and they die in the pool where they were born.

If we accept the evidence that these salmon are killing themselves with glucocorticoids, we still must wonder why. A venerable theory states that the fish are fertilizing the streams where their eggs will hatch, adding nutrients to the ecosystem that nourishes the insects on which the young will feed. But although ecologists have been able to show that salmon carcasses are an important source of nitrogen and phosphorous in streams and pools where they die, it is still difficult to imagine how programmed death could evolve on that basis. The problem is that these chemical resources are widely shared. The benefit of a salmon's death is not limited to its own offspring, and standard evolutionary theory stipulates that the driving force behind natural selection is the benefit to one's own progeny.

Although in nature it's difficult to say how much of the damage that leads to post-reproductive death of salmon is due to a flood of sex hormones that increase egg production and how much is due to programmed death, controlled experiments suggest programmed death. First, if the adrenal tissue—source of the glucocorticoid hormone—is removed from the salmon *after she spawns*, the result is: she lives. Second, salmon in captivity go through all the sexual development leading to egg laying, but they don't have to fight their way upstream, and they don't lay eggs. Nevertheless, they die, and from the same glucocorticoid malady. Third, the Atlantic salmon, close cousin to the Pacific salmon, undertakes an equally arduous migration but lives to swim back downstream to the sea for a second or even third cycle of ocean feeding, growth, and reproduction.

Octopus Anorexia

Some organisms are genetically programmed not to eat after reproduction and starve as a result; it's quicker and surer than traditional aging. Mayflies entering adulthood have no mouth or digestive system whatever. Elephants chomp and grind so many stalks and leaves during a lifetime that they wear out six full sets of teeth. But when the sixth set is gone, they won't grow another, and the pachyderms starve to death.

Octopuses make an especially good story. They live a short time, a few months to a few years, depending on the species, and they die after reproducing once. The female guards and cares for her eggs, but if conditions are not right for her brood, she may eat them, and then she has another chance to try again later. If she decides the time is right to deliver her young, not only does she refrain from eating her eggs, she stops eating altogether. Horton the faithful elephant in Dr. Seuss's classic children's book ("I meant what I said and I said what I meant") had nothing on the octopus mom who guards her eggs from predators, focused and immobile for months on end. During this time, her mouth seals over. She may live for years in this state of suspended animation, just guarding her eggs; but when the eggs hatch, she dies within a few days. Her death isn't from starvation. We know because there are two endocrine glands, called "optic glands" (though they are unrelated to the eyes) whose secretions control mating behavior, maternal care, and death. The optic glands can be surgically removed, and the octopus mom lives longer. If just one optic gland is removed, the female doesn't eat but still lives an extra six weeks. If both optic glands are removed, then the octopus doesn't lose her mouth and resumes eating after the eggs hatch. She then regains strength and size and can live up to forty weeks more.

Octopuses have a far more sophisticated intelligence than any other invertebrate. They play, and they learn. They can plan and strategize and solve problems. With light sensors on their tentacles and skin, they sense the background color and texture behind them, and some recolor themselves instantly in a pattern indistinguishable from whatever is behind them. But for all their intelligence, they are not easily bored.

In 2007, Bruce Robison of the Monterey Bay Aquarium Research Institute discovered a deep-sea octopus mom watching over her clutch of 160 eggs in the deep, cold waters off the California coast. He returned periodically to observe the same octopus on the same rock in the same position. From 2007 to 2011, she didn't eat, and she didn't move except to slowly circulate the water over the eggs, assuring a fresh supply of mineral nutrients. After four and a half

years, the eggs hatched, and the octopus mom disappeared, presumed dead, all within a few days. The empty eggshells were observed, memorializing her effort. It was the longest gestation ever observed, surpassing that of the frilled shark (*Chlamydoselachus anguineus*) by nearly a year.

Longevity Records Belong to the Plants

In 2014, photographer Rachel Sussman published a coffee-table volume of her ancient subjects entitled *The Oldest Living Things in the World*. All of them are plants. One reason for this, at least in comparison to ambulatory animals, is that plants don't have to worry about leg muscles strong enough to walk. Confined to one location, they can grow larger and sturdier, older and far more fertile than any animal, and reap the benefits of seniority.

Plants have another longevity secret. Early in the life of a developing animal, the sex cells, or *germ line*, segregate from the rest of the body, or *soma*. Only the germ line must be preserved immaculate to become the next generation; the body can afford to be sloppier with cells of the soma and take shortcuts as they reproduce themselves. But plants have a different system. The germ line and soma never really segregate. Plants, like animals, have stem cells, and in plants those stem cells give rise not only to new plant growth but also to the seeds and pollen that are destined to become the next generation. In a tree, the stem cells are located in a thin layer under the bark, called the *meristem*. The meristem extends into every branch and twig of the tree and gives rise to new leaves and also to buds and seeds. In some ginkgoes, nonflowering trees that date back to the Permian 270 million years ago, this meristem lineage may continue reproducing clonally (without sex) for millions of years.

The point of all this is that an animal can afford to let its stem cells gradually deteriorate, since they are destined to die with the individual organism; but a plant's future legacy is embodied in its meristem, and these cells must not be permitted to degrade with age.

There is a deep evolutionary reason why non-aging is so much more common in plants than in animals, and it is at the center of my theory, the Demographic Theory of Aging, which will be described in chapter 7, and elaborated in chapter 8. Aging evolved to help prevent famines, and only animals but not plants have to worry about famine.

The record-holder in Sussman's book is the Pando Grove, 106 acres of aspen in Utah grown from a single seed and still sporting a single root system. It is eighty thousand years old. All the entries over ten thousand years are root systems with renewing tops, like the aspen. The oldest single tree is an Antarctic beech tree in Tasmania, clocked at six thousand years. California sequoia trees are impressive for their size, but the oldest of them is less than three thousand years old. The only animal in the volume is a two thousand–year-old brain coral off Trinidad. Similar to the Pando Grove, it has grown from a single egg but may be considered a colony. Though it is an animal by genealogy, coral doesn't bother about locomotion and so is not subject to the size limits of mobile animals and especially land animals. Other corals collected off the Hawaiian Islands are reportedly more than four thousand years old, though they grow very slowly and never get very large.

Do Trees Age?

Some do, and some don't. Of course, size itself becomes a hazard as a tree becomes the tallest in its grove—the first to be struck by lightning, the most top-heavy and vulnerable to toppling in the wind when erosion weakens the roots' hold on terra firma. But in addition to this, it seems that most trees have a characteristic age, after which death finally becomes more likely with each passing year. Shoots ("epicormic sprouts") begin to grow directly from the tree trunk as growth at the outermost branches slows. There is some indication that trees become more vulnerable to fungus and disease with old age, but for the most part, old trees succumb to the mechanical hazards of excess size. The very ability to continue growing that offers them the possibility of "reverse aging" over so many decades proves in the end to be their downfall.

Later we will look at the need for population turnover, so that species can adapt and change over long periods of time. Most young animals and plants have trouble competing with elders that are larger, better established, less vulnerable to just about every threat in nature. But trees have this problem in spades. What chance for survival has the mere seedling, shaded by the mighty oak that is monopolizing the sunlight hundreds of feet above? Perhaps trees are evolved eventually to die so that their seedlings have a chance to start anew with a varied genome. (This theory is fleshed out in chapter 8.)

Aging in Reverse

In 1905, the Dutch biologist F. Stoppenbrink was studying the life cycles of *Planaria*, a kind of flatworm, a fraction of an inch long, common in freshwater ponds. He noted that when the animals didn't have enough to eat, they systematically consumed themselves, beginning with the most expendable organs (sex), proceeding to the digestive system (not much use in a famine), and then muscles. The worms got smaller and smaller until the most precious part—the brain and nerve cells—were all that remained. Stoppenbrink reported that when he started to feed the worms again, they grew back, rapidly regenerating everything they had lost. What's more, they looked and acted like young worms, and when their cohorts who had not been starved began to die of old age, the starved-and-regrown worms were still alive and kicking. This trick could be performed again and again. As long as Stoppenbrink kept starving and refeeding the worms, they went on living without apparent signs of age.

The immortal medusoid *Turritopsis nutricula* achieved its fifteen minutes of fame when it was hailed as "the immortal jellyfish" in science news articles of 2010. The adult *Turritopsis* has inherited a neat trick: after spawning its polyps, it regresses back to a polyp, beginning its life anew. This is accomplished by turning adult cells back into stem cells, going against the usual developmental direction from stem cells to differentiated cells—in essence driving backward down a one-way developmental street. Headlines called *Turritopsis* the "Benjamin Button of the Sea."

Carrion beetles (*Trogoderma glabrum*) perform a similar trick, but only when starved. As they play life out on a carcass in the woods, the beetles go through six different larval stages in succession, looking like a grub, and then a millipede, and then a water glider before ending up as a six-legged beetle. A pair of entomologists working at the University of Wisconsin in 1972 isolated the sixth-stage larvae (when they were just ready to become adults) in test tubes and discovered that without food, they regressed to stage-five larvae. If they were deprived of food for many days, they would actually shrink and regress backward through the stages until they looked like newly hatched maggots. Then, if feeding was resumed, they would go forward again through the developmental stages and become adults with normal life spans. They found they were able to repeat the cycle over and over again, allowing them to grow to stage six and then starving them back down to stage one, thereby extending their life spans from eight weeks to more than two years.

Ancient Aging

Hydras are radially symmetrical invertebrates, each with a mouth on a stalk, surrounded by tentacles, which grow back when cut off—like the many-headed monster of Greek mythology for which they are named. With their tentacles, they snare "water fleas" and other tiny crustaceans on which they feed. Some hydras are green, fed from the inside by symbiotic algae living beneath their translucent skin.

Hydras have been studied for four years at a time, starting with specimens of various ages collected in the wild, and they don't seem to die on their own or to become more vulnerable to predators or disease. In the human body, certain cells, such as blood cells, skin, and those of the stomach lining, slough off and regenerate continuously. The hydra's whole body is like this, regenerating itself from stem cell bedrock every few days. Some cells slough off and die; others, when large enough, grow into hydra clones that bud from the stalk-body to strike out on their own. This is an ancient style of reproduction, making do without sex. For the hydra, sex is optional—an occasional indulgence.

One recent article claims that the hydra does indeed grow older, and it shows it by slowing its rate of cloning. The author suggests that perhaps clones inherit their parents' age. The hypothesis is that only sexual reproduction resets the aging clock. If this is true, then the hydra's style of aging is a throwback to protists, ancestral microbes more complex than bacteria. Amoebas and microbes of the genus *Paramecium* provide examples of protists, a vast lineage that has anciently radiated into over one hundred thousand species and includes all the seaweeds, slime molds, ciliates, and other organisms. Their style of aging will be detailed in chapter 5.

Bees That Can Turn Aging Off

Queen bees and worker bees have the same genes but very different life spans. In the case of the queen bee, *royal jelly* switches off aging. When a new hive begins, nurse bees select one larva to be feted with the liquid diet of royalty. Some physiologically active chemical ambrosia in the royal jelly triggers the lucky bee to grow into a queen instead of a worker. The royal jelly confers upon the queen the overdeveloped gonads that give her a distinctive size and shape. The queen makes one flight at the beginning of her career, during which she might mate with a dozen different drones, storing their sperm for years to come.

Weighted down with eggs and too heavy to fly, the full-grown queen becomes a reproducing machine: she lays at a prodigious rate of about two thousand eggs per day, more than her entire body weight. Of course, such reproductive regality requires a suite of specialized workers to feed her, remove her waste, and transmit her pheromones (chemical signals) to the rest of the hive.

Worker bees live but a few weeks and then die of old age. And they don't just wear out from broken body parts. We know this because their survival follows a familiar mathematical form, called the Gompertz Curve, which is a well-known signature of biological aging. Meanwhile, queen bees, though their genes are identical to those of the workers, show no symptoms of senescence. They can live and lay for years and sometimes, if the hive is healthy and stable, for decades. They are ageless wonders. The queen dies only after running out of the sperm she received during her nuptial flight. At that point, she may continue to lay eggs, but they come out unfertilized and can only grow into stingless drones. Then, the same workers that formerly attended her assassinate the depleted queen. They swarm about her, stinging her to death.

Post-Reproductive Life Spans

Why is there a menopause? This has long been recognized as an evolutionary conundrum, to which theorists have supplied an answer that doesn't stand up to scrutiny.

In the charts on page 67, there is a dark line for fertility that is plotted along with the light line for survival. Neo-Darwinian theory, with its assumption that nature wants nothing but surviving, reproducing individuals, says that the dark line should continue throughout the life span. But there are human females in the upper left-hand corner, with fertility that soars in their teens, peaks in their twenties, and then vanishes in their forties—and they go right on living after the dark line has dropped to zero.

Neo-Darwinian theory says that to go on living after fertility has ended is a misallocation of resources. The individual gains no evolutionary advantage from continuing to live when it can no longer reproduce. Any resources spent in maintaining the body after fertility has ended are wasted. Natural selection deals harshly with such costly mistakes. That's why we weren't totally surprised in the case of the semelparous animals and plants above to learn that they all die promptly after reproduction is finished.

But people are not salmon. We care for our young and our extended families, and our devotion continues after our children have grown and become parents themselves. Hence, the standard explanation for life that continues after fertility ends is called the "grandmother hypothesis." Women have a genetic interest in seeing their grandchildren grow up healthy. Maybe at age sixty they can contribute more to their own legacy by caring for their grandchildren than by having more babies of their own. This is a hypothesis that sounds reasonable, at least for humans, but demographic researchers have found that when they do the numbers, it's hard to make it work.

It gets worse. When we compare aging among animals quantitatively à la Baudisch, we find many animals whose fertility falls off before their survival. Whales and elephants are examples of organisms that outlive their fertility. They are social animals, too. Perhaps they are more important to their grandchildren than we know. But in the chart, there are other animals that go right on living after their fertility has ended. These include guppies, water fleas, roundworms, and bdelloid rotifers, all of which make deadbeat dads look like Mary Poppins. All these animals lay eggs, and that's it. None of them lifts a wing or a fin to care for their young, let alone their grandchildren. Yet modern evolutionary theory says that there is no natural selection to keep them alive, and thus we should expect them to be kaput.

In 2011, Charles Goodnight and I had an idea about how post-reproductive life span might evolve, an idea that sounds pretty unlikely in the abstract, but when we did the numbers, it actually panned out. An older, "retired" segment of the population, we argued, serves to keep the population stable over cycles of feast and famine. When times are good, they eat the excess food and help prevent population overshoot. When food is scarce, they are the first to die. I will return to this idea to flesh it out when we have more context in chapter 7.

Rotifers and Dandelions

Evolutionary biologists have long been intrigued by sex. Why is it that most animals and plants don't reproduce the most obvious and efficient way—by cloning? Why do they willingly sacrifice half their genes (and half their fitness!) and subject themselves to a popularity contest in order to join with a partner and reproduce in this clumsy, inefficient, relatively precarious mode?

There are several competing theories, but all—couched under the general name the Red Queen—agree that sex maintains diversity, and this must be important, nay, indispensable. How else to explain that sex has managed not just to survive but to prevail in the face of competition from the simple efficiency of clonal reproduction?

Bacteria have their own form of sex, sharing genes in a way that often has nothing directly to do with reproduction. Plants that have roots that propagate underground (a form of cloning) also produce flowers and seeds occasionally, presumably to mix their genes. Almost all animals have sex of one sort or another, though some have a choice of reproducing with or without sex. Earthworms fertilize each other's eggs. Roundworms fertilize their own eggs with their own sperm, but one in a thousand is a male, and this tiny minority is enough to maintain diversity.

In this context, the bdelloid rotifer is something of an evolutionary scandal. No males have been found in any of its 350-odd species. Although their ancestors may well have been sexual, today these microscopic animals don't require males to reproduce. Often prey to fungal parasites, bdelloid rotifers can dry out to elude them, resuscitating in a drop of water: such behavior may compensate for the variety they lack from not having sex. Dandelions, clonal plants that we've all seen, are also a bit of a mystery from the vantage point of their persistent *a*sexual reproduction; however, studies show that they somehow maintain genetic diversity despite their lack of sex.

Instant Replay

Styles of aging in nature are just about as diverse as they can be—from mayflies that live a day to turtles that show no aging and trees that live thousands of years. The diversity is not just in the length of life but also the manner of death and the contour of the aging curve. All this variety suggests that nature is able to turn aging on and off at will. With this in mind, we may be forgiven for regarding theories that explain why aging must exist with extreme skepticism. *Whatever* our theory of aging turns out to be, it had better make room for plasticity, diversity, and exceptions. Neo-Darwinian theorists claim that nature does its best to avoid aging but that the variety of genes available for natural selection limits her options, and it is the limitation that makes aging inevitable. On its face, this seems like a dubious claim. Looking at the panoramic sweep of aging modalities, we find not limitations but variations as far as the eye can see.

THREE

Darwin in a Straitjacket: Tracing Modern Evolutionary Theory

It doesn't matter how beautiful your theory is, it doesn't matter how smart you are. If it doesn't agree with experiment, it's wrong.

—Richard P. Feynman, 1965 Nobel Prize in physics

Biology got its main success by the importation of physicists that came into the field not knowing any biology . . . The most important thing today is for young people to . . . actually know how to formulate an idea and how to work on it . . . I think you can only foster that by having sort of deviant studies. That is, you go on and do something really different . . . But today there is no way to do this without money. That's the difficulty. In order to do science you have to have it supported. The supporters now, the bureaucrats of science, do not wish to take any risks . . . Even God wouldn't get a grant today because somebody on the committee would say, oh those were very interesting experiments (creating the universe), but they've never been repeated. And then someone else would say, yes and he did it a long time ago, what's he done recently? And a third would say, to top it all, he published it all in an un-refereed journal (The Bible).

—Sydney Brenner, 2002 Nobel Prize for work on genetic code

When a distinguished but elderly scientist states that something is possible, he is almost certainly right. When he states that something is impossible, he is very probably wrong.

—Arthur C. Clarke

How Can People Believe Such Things?

This chapter is not necessary to the logic of the book. We included it for readers who may be wondering about our accusation that evolutionary science has taken a major wrong turn. How could it be that so many smart people were drawn into a scientific cul-de-sac? We found these profiles from the sociology of science to be fascinating in their own right.

During the twentieth century, the legacy of Charles Darwin was hijacked. The name of one who devoted his life to observing nature came to be attached to a theory that put mathematics first and stripped the biological world of its rich complexity in order to force it into an inept framework. Substituting for fieldwork and natural history, laboratory breeding experiments were offered as proof that all mechanisms of evolution conform to the narrow principle of one gene at a time. This is the context in which our present understanding of aging was molded, and it will be necessary to break the spell of mathematical hubris before we can peel back the theory to see aging for what it really is.

Indeed, to the happily unschooled, the current evolutionary theory of aging seems rather bizarre. Many people who hear it for the first time ask incredulously, "Is that *really* what they think?" The essence of the mainstream theory is that natural selection, although rewarding the fastest reproducer, is stuck with ineluctable biochemical constraints that link rapid growth and reproduction to dire consequences decades after the fact. A genetic connection links your fertility at age twenty with your dementia at age eighty, and it is so close that, try as she might,

Mother Nature has never been able to pry the two apart. To understand where this notion comes from, and why it has become broadly accepted among scholars who have mastered the mathematical theory of evolution, we need to look back at the history of evolutionary thought in the half century after Darwin.

In the early years of the twentieth century, some scientists faced up to the fact that Darwin's Theory of Evolution was not really a theory at all. Not in the modern sense. They took on the project of interpreting and clarifying evolution as a deductive logical system. Right or wrong, their system was far more explicit and therefore *testable* than Darwin's version. The foundation of evolutionary theory was laid by people whose background was more mathematical than biological, with consequences that remain imprinted on the science as it is practiced today.

For scientific theories, testable predictions require validation by experiment. Darwin's account of evolution was a way of understanding the history of life. It supplied a meaning, a context, a story of how life came to be the way it is today. It could be evaluated for plausibility. How well does it fit with what we know about the way the world works? But how do we test Darwin's theory? The smallest experiment I can imagine would require an island the size of Madagascar and about fifty thousand years. Stuck as usual in its budget constraints and myopic perspective, the National Science Foundation has stubbornly refused to set aside a paltry trillion dollars for such a project.

To understand what these turn-of-the-century scientists were aiming at and what they found lacking in Darwin, let's take a trip back in time to visit Darwin and a Bavarian friar who was Darwin's contemporary.

Evolutionary biology is not the only science that suffers from an inability to do experiments. Astronomy has similar problems, requiring yet larger scales and longer times than evolution. And human epidemiology has problems of an ethical sort: you can't

very well dictate people's diet or behaviors as if they were lab rats on a thirty-year experimental run.

But evolutionary biology today is a uniquely sick science, missing the vibrancy, the audacity, and the commitment to empirical truth that form the core of the scientific method. Many prominent biologists and several Nobelists have called for a rethinking of evolution's foundation, but this is not happening, not yet.

There are at least four reasons why evolutionary biology is in trouble today.

- From its beginnings, Darwin's ideas were hijacked by British social Darwinists who sought to contort them into a justification for class privilege. The imprint of social Darwinism remains as a distortion to evolutionary thought.

- In the early twentieth century, the core theoretical principles of neo-Darwinism were created by mathematical scientists who didn't know much biology. Even today, the field remains divided between scientists who are adept at mathematical theory and scientists who are intimately familiar with natural ecologies. The two wings of evolutionary science don't talk to each other enough.

- Natural selection cannot be observed in the wild, because it requires huge areas and thousands of years. Lab scientists have substituted breeding experiments to give the science a veneer of empiricism; but the breeding experiments have been designed to replicate conditions of standard theory, not conditions of nature.

- Especially in America, evolutionary science is under siege by Christian fundamentalists, who wish to preserve the sanctity of the biblical account of creation. The scientific community has responded by circling its wagons, attacking every criticism of the theory as if it were faith-based nonsense.

Darwin Was Afraid of Sex; He Shouldn't Have Been

In fact, sex might have helped him with his biggest problem—but there was no way he could have known that. Not without advice from a celibate monk named Gregor Mendel. Allow me to explain.

Darwin's theory begins with the fate of different individuals and their progeny. Some individuals survive and reproduce more than others. Darwin realized that, in a highly competitive biological environment, these little differences in success would accumulate over the generations. Over a long course of time, they could make major changes, including whole new life-forms. Such incremental changes could lead, Darwin reasoned, to the evolution of new species.

However, Darwin's theory depended on these differences being inherited, and exactly how they were inherited wasn't known to Darwin. There was an art of breeding plants and domestic animals, already thousands of years old. Dogs were bred to make their snouts longer, chickens to lay more eggs. While it was apparent that offspring tended to resemble their progenitors, the details were cloudy.

Darwin knew that a tall person who married a short person was likely to have children of an in-between size. If this were the general rule, then it would create a big problem for his theory, because the theory depended on having a diverse population of individuals in order for natural selection to operate. If everyone is the same, then no one has a particular advantage, and there is nothing for natural selection to select.

Now just imagine a world where every individual was in every way the average of her two parents. The extremes of the population would tend to be leveled out just by random pairings. Long-beaked birds and short-beaked birds would all tend to have baby birds with medium beaks. White rabbits and brown rabbits would have bunny babies that were tan. Cheetah cubs with one especially fast parent and one sort-of-slow parent would tend to a middling speed. How many generations would it be before individual differences were erased? How long before all diversity in an area would be diluted out of existence? Natural selection would have nothing to do in Lake Wobegon, "where all the women are strong, all the men are good-looking, and all the children are above average."

If everyone were the same, then evolution would grind to a halt. Natural selection feeds on diversity. Without variation, there could be no evolution. If sexual reproduction tends to average out all the differences, then natural selection could not work. This was a big problem for Darwin's theory, and he knew it. He worried about it his whole life.

At the same time that Darwin was writing his magnum opus, six hundred miles to the east in a friary in the Bavarian mountains, a bespectacled, wonkish monk was experimenting with pea plants. Gregor Mendel fertilized a tall pea plant with pollen from a short pea plant, and the seeds all grew tall—every one of them. The tall/short hybrids were all just as tall as the purebred tall plants! That must have astonished him.

He traced plants of the next generation, grown from the tall/short hybrids pollinating one another, and found that three-fourths of them were tall, and one-fourth were short. Well, actually, it wasn't as quick or as easy as this. It took six years of trying combinations, purifying purebreds, hybridizing hybrids—and then hybridizing the purebreds and purifying the hybrids. He had the benefit of the simple monk's life, protected from distractions and responsibilities with bodily needs provided. In the spring and summer, he worked meticulously to group and isolate the plants to control the pollination process. In the summer and fall, he scrutinized the plants one by one, characterizing their shape, color, and size, recording everything. In the winter months, he had the work that a scientist lives for: slicing and dicing the data in all different ways, looking for patterns, trying out hypotheses, becoming excited and disappointed in turn as each idea passed through its cycle of trial and failure. Between 1857 and 1863, Mendel grew twenty-nine thousand pea plants, classified them, charted them, counted and recounted and calculated ratios and probabilities. Besides the plants' height, he traced six other characteristics (color, shape, etc.) and found them to be independently inherited, with all combinations represented. By 1865, he had the outline of an inheritance scheme that must have seemed to him so

abstract and gamelike that he could hardly imagine that real, living things behaved in such a mathematically regular way.

This was fifty years before the word "gene" was invented and one hundred years before genes came to be identified with the ladderlike, self-replicating molecule, DNA. But Mendel already had the abstract idea that there were particles of inheritance—Mendel called them "*faktorem*," factors—that determine all the individual's various traits. Every pea plant has two height factors, and a tall factor will trump a short one. Only plants with two short factors actually appear short. These two factors came from the plant's seed and pollen, its mother and father. One of these factors is chosen at random and passed to the next generation. It can be the factor that came from the mother or the father, with equal likelihood. Every individual carries one of its father's two factors and one of its mother's.

This was the system that Mendel had decoded. In 1865, he presented his findings to the Natural History Society of Brno (currently in the Czech Republic). He wrote and published a paper the next year. Later, after publication of *On the Origin of Species* had made Darwin famous, Mendel sent a copy with a letter of explanation to Darwin, which Darwin neglected and never opened. Because of this accident of history, Darwin's science languished for forty years.

In Mendel's factors was the resolution to Darwin's biggest fear, the collapse of diversity. Sex shuffles the factors but does not change or soften or dilute them. Any averaging is done at the level of appearance only, but the potential for every trait (and the extremes of each trait) remains intact in the population. Far from collapsing diversity, sex has the power to mix and match different combinations of traits. Mendel's factor game assures that no two individuals are alike. Sex is the friend of diversity. It is also worth noting, since this book is about the power of groups in evolution, that obligate sexual reproduction—the imperative to share genes—predisposes species to become more social, since they must pay attention to potential mates; sexual reproduction is a preadaptation for social evolution.

A Modern View of Sex

Joel Peck at the Department of Genetics, University of Cambridge, takes a less Victorian view. He just loves sex. To Peck, sex represents nature's solution of how to promote cooperation, how to tie communities together and prevent infighting. It is clear that natural selection is in danger of devolving into a dog-eat-dog battle for king of the heap.

Sexual sharing of genes is an impressively complex mechanism, a "masterpiece of nature" as Graham Bell titled his book on the subject. Explaining how it came about is one of the greatest challenges to confront evolutionary biology, and it is widely recognized as such. Some researchers have devoted their lives to the subject. Others have thrown up their hands and taken sex as a given so they can proceed with other studies.

But what is clear is this: the sharing of genes has enormous advantages for a "deme" (this is the word for an intrabreeding community). Sex makes possible a level of cooperation that could never evolve without it, and tightly cooperating communities are devastating competitors compared to any ragtag bunch of backstabbing individuals.

So it may not yet be possible to say for sure how sex arose. (Life managed its first two billion years without the sort of sperm-egg cell merger type of sex—called meiotic sex—necessary for the existence of us and most other animals and plants.) But why gene-sharing communities have beaten out the competition (communities that don't have sex) has a more straightforward explanation. In sexual communities, a great deal of the temptation to behave selfishly has been removed, because no gene can spread unless it works well in combination with others. Sex has played an important role in the process of integrating groups into effective, cooperating units.

A Twentieth-Century Science: Origins of Neo-Darwinism

If Darwin didn't know about Mendel's theory, no one did. Mendel's solution to the schema of genetic inheritance was hiding in plain sight, while evolutionary thinkers wandered in the desert for forty years. Then, in 1900, Mendel's work was rediscovered, and an opportunity opened to join his work with Darwin's. Could Darwin's descriptive theory be made into a quantitative, predictive science? The challenge was taken up by several mathematical scientists, working independently and corresponding with each other over the first decades of the twentieth century: Alfred Lotka, Sewall Wright, J. B. S. Haldane, Theodosius Dobzhansky, and foremost, R. A. Fisher.

Imagine yourself inside the brain of Ronald Fisher, mathematically inclined with an extraordinary facility for abstraction, passionately devoted to Darwin's ideas, turning them around and around in a quest for a more explicit, quantifiable, and predictive science. Darwin painted a picture of a struggle for existence, "survival of the fittest."* So the first step was to be able to measure "fitness" and attach a number to it. This was not so hard to do. Fitness in Darwin's world consisted in producing more offspring, faster. Fisher borrowed from Lotka a formula that counts offspring, with a bonus for those produced earlier in the life cycle. That was a good start.

But the next step was not so easy. "Fitness of what?" was the question. You could say it was the fitness of the individual. That would work in a population without sex, where each organism reproduced clonally, and you might track the growing number of copies of the most successful variety. But in a sexual population, no two individuals are alike. Fisher wanted a mathematical theory of how populations change over time. He wanted something he could count and track over many generations, and individuals were too ephemeral for this purpose. Individuals have fitness, but individuals don't evolve. Populations evolve as

* Darwin's colleague and promoter Herbert Spencer coined the phrase "survival of the fittest," and Darwin himself picked it up in later editions of *The Origin*.

their compositions change, but populations don't have "fitness." This was the dilemma Fisher faced.

And here is the solution that Fisher's fertile forebrain concocted: he would attach "fitness" not to an animal or plant but to each gene within it. The genes, then, persist over time, and a measure of their success is their prevalence in the population. At any given time, some individuals have the gene and some don't. Count how many copies of a gene exist in a population, and that is a measure of the success of the gene. A gene that is succeeding grows in prevalence from one generation to the next. More individuals have that gene (compared to an alternative, slightly different version).*

Each gene makes a contribution to the success of its bearer in surviving and proliferating. The gene is rewarded when more copies of it appear in the next generation. The reward accumulates over time in a process that can be charted quantitatively. Some genes will spread through the population, while others will languish and ultimately disappear. This is Fisher's quantitative model for Darwinian evolution. Fisher gave to Darwin's theory something that could be measured and calculated, predicted and tested.

This was a brazen intellectual leap, the birth of the selfish gene, half a century before Richard Dawkins would immortalize the phrase. This way of thinking, this model of reality, makes it possible to develop a mathematical theory of evolution—*but* it glosses over vast areas of biological reality. In the 1920s, Fisher's detractors pointed out that the concept of a gene's fitness is problematic. Fitness depends on all the genes in an individual working together in a harmonious way. (Genes for thicker bones would be an advantage in muscular individuals who carry extra weight, but in light and nimble individuals, they might just slow them down.) For that matter, the fitness of an individual depends entirely on ecological context. (The thick fur of the polar bear is an asset in snowy, northern climes but a detriment in the tropics.)

Fisher replied that the world is a big place and everything averages out. In the long run, each gene would have to work with different com-

* Different versions of the same gene are competing. A version of a gene is called an "allele," and technically we should be talking about the prevalence of an allele, rather than the prevalence of a gene. In using the phrase "gene prevalence," we have chosen the more informal usage. "The Selfish Allele" lacks that certain *je ne sais quoi*.

binations of genes as they appear in different individuals and with different individuals as they are confronted with different life situations in different environments. It would be the "generalist genes" that would do best overall, those genes that could make a positive contribution to fitness no matter what the circumstance.

But what about the tendency we see in nature for every opportunity to be seized by a specialist, able to defend his little bit of turf because no one knows his particular business as well as he does? For example, in arid, hot regions around the world, there are about nine hundred separate species of fig. And every fig species is pollinated by its own specialized wasp—nine hundred kinds of wasp that have evolved to consume the nectar from one fig species and one only.

Through his career, Fisher's counterpoint was Sewall Wright, who kept trying to put the "fit" back in fitness. He wrote about relationships and combinations, which gene works better with which other gene, and what kind of variety might be better adapted to one environment than another. Wright had the longevity advantage, and he continued thinking and writing up until his death in 1988 at age ninety-eight. Fisher (1890–1962) was long dead by that time, but it mattered little because Fisher had prevailed in the debate and won the community over to his perspective.

However, in retrospect, Fisher made some assumptions that turn out both to be wrong and to color our understanding of evolution. These, in my view, are the major weaknesses of Fisher's model:

- Fisher assumed that genes contribute independently to fitness, but in fact, genes interact strongly.
- Fisher assumed that mating is at random, when in fact, mating choices are strongly based on geography and instincts about compatibility.
- Fisher assumed that total population sizes are static, when in fact, populations fluctuate. (The fact that populations can "fluctuate" and have fluctuated to zero—becoming extinct—will turn out to be crucial in our new understanding of aging.)

- Fisher assumed that ecosystems present a static background, when in fact, ecosystems change, critically affecting their inhabitants.
- Fisher assumed that local environmental variations all average out, but in fact, species adapt exquisitely to local variations.

You might imagine that with such a flawed foundation, Fisher's theory should be completely discredited, but there are actually good reasons to give a theory a chance to prove itself, even if the assumptions that go into it are oversimple. Most scientific theories have logical problems at some level, but still there are large areas where they describe reality in a useful way. So let's give Fisher's theory a chance, even though it may skip over some major properties of life, and see whether its predictions match what nature has served up.

"Neo-Darwinism" is the shorthand we've used in this book for Fisher's version of evolutionary theory. The fact that its foundational postulates are demonstrably wrong is not necessarily a fatal flaw. But to be sure, we need to keep our eyes open and see how well the predictions of neo-Darwinism correspond to what we actually find in nature. We might expect that scientists would be wary of such a theory and that they would retain a healthy skepticism in evaluating its predictions.

But that's not how it happened. Ultimately, Fisher's theory succeeded because it created a fertile, if insular, intellectual universe. Working out the mathematical consequences of the selfish gene provided gainful employment and an interesting occupation for hundreds and then thousands of interested intellectuals, prematurely infatuated with the theoretical clarity Fisher's mathematical marriage of Mendel and Darwin achieved.

Our bodies have ten times as many bacterial cells as human cells, and 90 percent of the genes in our bodies are in the microbiome. (Of course, the bacterial cells are much smaller, so they are not a

major part by mass.) University of Wisconsin medical microbiologist Margaret McFall-Ngai hypothesizes that our marine ancestors lived in an ocean so full of microbes that the immune system likely originated as a way not of keeping strangers out but of selectively welcoming them in.

Social Darwinism, a political ideology with racist and classist overtones, rationalizes oppression of the underclass as the natural right of the privileged. The ideological perversion of Darwin's ideas led to experiments in eugenics and national socialism, a.k.a. Nazism. One of the many fallacies of social Darwinism is the failure to recognize the value of diversity. Together, we are smarter and stronger and more robust than any of us, and a utopia of ideal clones would be no utopia at all.

Racial purity is a meme compromised by the reality of vast numbers of distinct organisms—including those that comprise the "superorganism" of our human form—that evolve together as bodies. An index of the "sacred impurity" of mixed assemblages can be gleaned from the fact that, in the laboratory, pure monocultures of one type of microbe often won't grow. Indeed, so many constituents of the human microbiome have defied isolation that biologists have begun to talk of microbial "dark matter." One half of microbes in the human body cannot live anywhere else. How, then, can they be studied in a laboratory?

Life is interdependent at every level, and human societies are most resilient when they are most diverse. The ascent of American culture and power in the world coincided with bloating of the melting pot. William Frey calls it the "diversity explosion," and it is turbocharging the engine of change, not just in America but the world over.

How Experiments in Lab Evolution Lent Credibility to Neo-Darwinism

Fisher's theory of one gene at a time does not accord with what we see in nature (rampant interdependence!). But it is difficult to falsify it absolutely, because we cannot design experiments for evolution to perform. We must be content to sleuth the fossil record instead. Substituting for experimental evolution, laboratory scientists have used breeding experiments to validate Fisher's theory. And indeed, the theory works like a charm in that context. In fact, a huge literature has been compiled, and almost all supports Fisher's version of reality.

The catch is that the lab experiments are all *designed* to select one trait at a time! When animals are bred for long hair or small size or high fertility or long life—whatever the trait—breeding can enhance it. But it is a mistake to take this as evidence that natural selection itself tends to select one trait or one gene at a time.

Theodosius Dobzhansky (1900–1975) was unique among the mathematical evolutionists who inaugurated the modern synthesis of neo-Darwinism in seeking experimental support for his ideas. As a young man in Ukraine, Dobzhansky studied statistics and advanced the early theory of population genetics. In the 1930s, after he emigrated to the United States, Dobzhansky set up a lab breeding fruit flies, and he used the experiments to corroborate predictions of neo-Darwinian theory. These results were taken as empirical support for the equations, and they lent the impression that the theory was on a sound experimental footing. But this reasoning was subtly circular. The selection experiments were built to emulate the kind of selection that comprised Fisher's narrowly limited vision. Populations were held constant by discarding most of each generation, while representatives of each new generation were selected in proportion to the fertility of the last generation.

The experiments were designed around the mathematics of neo-Darwinism, and though they are a realization of Fisher's mathematics, they do not bear at all on the most questionable aspect of his model: whether selection in nature acts in this way.

The title of Dobzhansky's essay, "Nothing in Biology Makes Sense Except in the Light of Evolution," has become a catchphrase in the field, because it expresses our sense that evolution is the source and raison d'être of all biological phenomena. It is curious to think that in Dobzhansky's own mind, the causal chain could be traced back one more step to the will of God—Dobzhansky remained a devout member of the Eastern Orthodox Church through his life, and he regarded evolution as the instrumentality of God's creation.

Science on Two Tracks

For thirty years, evolutionary science was proceeding on two parallel tracks, with different publications, different methodologies, and very little interaction. In one corner, you had Fisher and Wright duking it out, with theoretical implications being elaborated by J. B. S. Haldane and Theodosius Dobzhansky, both mathematically proficient biologists with an ability to write engagingly for professional as well as popular audiences.

In the other corner, there were naturalists doing what Darwin did, observing nature and writing about what they saw, speaking loosely of adaptive behaviors and traits with a general sense that natural selection was about survival and reproduction, but without perceiving any need for the sort of rigorous and consistent theory trail-blazed by Fisher. This is not at all to denigrate their achievements. Philosopher of science Alfred North Whitehead, himself a gifted mathematician and famous collaborator with ultrarationalist Bertrand Russell, traced good science to a combination of theory and evidence. Facts without a theory are merely a catalog, and they offer no basis for extending our understanding.

Theory without facts is mathematical and abstract, and however elegant and sophisticated it may be, by itself it is not "science." As Whitehead emphasized, lucid speculations provide the thrust that propels science forward, but if these are not tested regularly and thoroughly, there is a danger of detachment from reality.

J. B. S. "Jack" Haldane (1892–1964) is credited as one of the founders of the theoretical framework called "population genetics," or "neo-Darwinism." Of Scottish ancestry, Haldane was a precocious polymath and son of a physiologist who wrote a paper with his sister, left his organs to science, and experimented on himself, exposing himself to toxins. At the age of four, as blood was being wiped from his forehead, he asked the physician, "Is this oxyhemoglobin or carboxyhemoglobin?" Sir Peter Medawar called Haldane "the cleverest man I knew." Haldane was involved also in early origins-of-life speculations suggesting, more or less simultaneously with Alexander Oparin in Russia, that the ancient chemical environment of Earth, rich in hydrogen-containing compounds, could have spontaneously given rise to life.

Politically, Haldane was a Marxist, but he grew disenchanted with Lysenkoism promoted in communist Russia. A friend of Aldous Huxley and an accomplished writer, Haldane predicted test tube babies in his book *Daedalus*, and this influenced Huxley's fictional eugenic dystopia *Brave New World*. However, without Haldane, modern selfish gene theory may never have been developed. He presented it in germ form after being confronted with the "problem" of altruism, which doesn't make sense in a world of selfish evolutionary actors. When asked how far he'd go to save someone, the geneticist supposedly pondered a minute and then grabbed a nearby napkin and started calculating. "I would jump into a river to save two brothers, but not one," Haldane concluded. "Or to save eight cousins but not seven." What seems like altruism (on the part of an individual) is really selfishness in disguise (on the part

of his genes) because the gene may provide support to copies of itself in close relatives of the individual. This cynical view of altruism was developed into the formal mathematical theory of kin selection by William D. Hamilton, and then popularized by Richard Dawkins in the book *The Selfish Gene*.

In 1956, the mathematically gifted Haldane, an atheist, left University College London to join the Indian Statistical Institute in Calcutta. He said he went both for the climate and to find a more equitable community, friendly to his socialism.

Do Animals Regulate Their Populations?

V. C. Wynne-Edwards (1906–1997) was an old-school naturalist, a broadly educated scholar and eloquent writer, who collected a massive amount of evidence from nature in a book on population control that was at once the denouement of his lifework and the cause of his fall from grace. His thesis was that animals control their population density at a sustainable level, responding to available resources and limits. They do this by spreading themselves out, by limiting litter size (what we might call "birth control"), and by territorial cues that signal other individuals of their species to respect their property claims. Some beetle species turn to cannibalism when crowded, and though lemmings don't actually jump off cliffs, they do respond to crowding by exploration en masse, and many die in transit.

Was Wynne-Edwards right? We can only judge by looking at the diversity of evidence. No single example can be expected to provide a slam-dunk proof, but the facts collected by Wynne-Edwards in support of natural population control are manifold and compelling. The breadth of examples looks even more impressive fifty years on. Whales and elephants have very low mortality rates, and so they reproduce much less frequently than their physiology permits. Lions and tigers spend much less energy on reproduction than smaller cats, and Wynne-Edwards says that is because they live so much longer that they would overpopulate if they reproduced more frequently. Flies bred in jars will reach a limited density and then cease to lay eggs, even if plenty of food is provided. Beetles eat their young in conditions of crowding. Mice and other rodents respond to crowded cages by refusing to reproduce, even when they have plenty of food, becoming pugnaciously territorial. Although Wynne-Edwards spoke of density-dependent population control for "the good of the species" in a way that bridled the selfish gene assumptions of the neo-Darwinists, these examples are not lightly to be dismissed.

There are others. Game fish can be bred in tanks, and the number of fish remains remarkably constant whether a proportion of the fish is periodically harvested or their tank is left undisturbed. Long-lived birds—penguins, auks, condors, vultures, eagles, albatrosses—lay only one egg at a time, even when the physiological burden of producing an egg is trivial. In fact, if the one egg is lost or broken, the bird will replace it. You have to wonder: If the birds can lay two eggs so easily, what stops them from doubling up in Darwin's lottery?

In the strong territorial and hierarchical tendencies of our own species, Wynne-Edwards saw the same density-dependent controls. He cited his predecessor Alexander Carr-Saunders, who compiled his own book a generation earlier, limited to anthropological examples. Hunter-gatherer populations were stable for hundreds of thousands of years before agriculture, and Carr-Saunders lists the ways that overpopulation was avoided: fertility limits and abortion, warfare, and even infanticide found their place. This looked to Carr-Saunders like a program for sustainability rather than maximal reproduction.

In the context of neo-Darwinian theory, population control is just as impossible as programmed aging. But animals in the wild must have been playing hooky the day that theory was covered.

Wynne-Edwards offered a compelling barrage, six hundred pages of evidence for natural population control. Still, the book ultimately failed in its mission because, it was said, his arguments were fuzzy and lacked mathematical rigor. His book was slammed on theoretical grounds, and even though the field evidence in Wynne-Edwards's book was never refuted, it became highly unfashionable to talk about the evolution of population control.

The Hostile Takeover

A few years after Wynne-Edwards's magnum opus, a young mathematical biologist named George Christopher Williams (1926–2010) entered the field with a smart, well-reasoned treatise that took Wynne-Edwards's work as a foil.

Williams had recently completed his graduate work at the University of Michigan and was well aware of the two tracks in evolutionary biology. He was aware that many of the evolutionary explanations put forward by naturalists to help understand what they observed did not comport with the methodology of neo-Darwinism or the mathematics of the selfish gene. He called the field biologists to task for failures of rigor. Naturalists were in the habit of attributing to every phenomenon they reported some plausible-sounding selective advantage that would be invoked to explain how this phenomenon had evolved. But we should expect more from our biological theory than a good story. Biologists need to think in terms of explicit mechanisms, and, where possible, relate their observations to quantitative predictions of evolutionary theory. The methodology of neo-Darwinism provided a basis for computing such predictions.

Williams went further, pointing out that much of the fuzzy thinking ascribed to naturalists could be traced to their excessive comfort in thinking about *collective* fitness. According to theory (said Williams), natural selection happens to one individual at a time. Individual genes proliferate (or die out) much more rapidly than the world around them changes. Natural selection on groups is a comparatively slow and inefficient process. Where there is a choice between an individual advantage and a collective advantage, the individual advantage is expected almost always to carry the day.

Social Darwinism, Ronald Fisher, and Eugenics: The Cultural Context of Neo-Darwinism

Anthropologists and students of the history of science are fond of pointing out that scientists aren't as objective as they think but are deeply influenced by their social context, funding, and culture. There is a sociology of science, even a politics of science. Friedrich Engels saw Darwin's theory clearly in this context:

> The whole Darwinist teaching of the struggle for existence is simply a transference from society to living nature of Hobbes's doctrine of *bellum omnium contra omnes* [the war of all against all] and of the bourgeois doctrine of competition together with Malthus's theory of population [the thesis that populations tend to grow far faster than their resources, imperiling them]. When this conjurer's trick has been performed . . . the same theories are transferred back again from organic nature into history and it is now claimed that their validity as eternal *laws* of human society has been proved. The puerility of this proceeding is so obvious that not a word need be said about it.

The full title of Darwin's 1859 book was *On the Origin of Species by Means of Natural Selection, or the Preservation of Favoured Races in the Struggle for Life*. After its publication, the British upper class lost no time drawing from it a scientific justification for class privilege. Thus social Darwinism was born.

At the end of the nineteenth century, the *eugenics* movement came out of this tradition. It was Darwin's cousin Sir Francis Galton who invented the term, which he defined as "the study of all agencies under human control which can improve or impair the racial quality of future generations." Troubled by the fact that "the rich get richer and the poor get children," the social liberals of the day were looking for humane ways to keep the gene pool of human genius from being diluted into extinction.

It was an unquestioned assumption that poor people were lazy and stupid, even if no one claimed that all rich people were geniuses.

If pigeons and dogs could be bred, so could people. But early proponents of eugenics should have been more circumspect concerning the dangers of inbreeding, which were already well known to animal breeders. Galton collaborated with the Wedgewood, Darwin, and Huxley families in an experiment of race improvement. They would only reproduce with one another, artificially selecting for a superior race. But the result of this bold four-family experiment was that within just two generations, most of their offspring either perished during birth or were born seriously handicapped.

Ronald A. Fisher was a genius of rare stature, not only the architect of neo-Darwinism but also the father of modern statistics. But Fisher's first passion was eugenics, and he spoke about the threat to the human gene pool with a passion that infused all his other work. Fisher's magnum opus was a 1930 book called *The Genetical Theory of Natural Selection*. Part 1 of his book is a standard reference for evolutionary scientists, justifying the assumptions and deriving the mathematical machinery at the core of evolutionary science as it has been practiced ever since. Part 2 of the book is a political screed on eugenics that can only be described as an embarrassment to readers of the first half.

Within a decade of the publication of Fisher's book, Adolf Hitler turned "eugenics" into a dirty word, a verboten idea that can no longer be discussed in polite company. But practitioners of the science rarely remember the fact that modern evolutionary theory grew from roots in a disgraced social philosophy. Much of the mathematical machinery of modern statistical analysis was developed almost incidentally, in service to a new quantitative science of evolution that supported the politics of genetic manipulation.

The Group Selection Debate

Following the books of Wynne-Edwards and George Williams, a debate ensued in the scientific literature, continuing through the mid-1970s. Williams and the smart and prolific British evolutionist John Maynard Smith argued against group selection. David Wilson recently and Michael Gilpin earlier each wrote a carefully argued book that sought to lend theoretical support to the idea of group selection, but neither Wilson nor Gilpin could overcome the prejudice that regarded group selection as fuzzy thinking, lacking in rigor.

Granted, the two worlds of evolutionary biology were long overdue for a session of truth and reconciliation. But in a more strictly scientific world, the naturalists, being closer to empirical reality, should have dictated to the mathematicians and not vice versa. What happened instead was that the naturalists were intimidated by mathematics, and they took the theorists at their word, lest they be regarded as stupid. The naturalists caved, the mathematicians triumphed. Soon every new biology student was being taught that group selection is a no-no. In the scientific literature, any appeal to group selection was taken as evidence of flawed reasoning. Any manuscript submitted to a journal could and would be rejected for such a lapse. Only slowly is this veil of censorship being lifted.

It is in such an atmosphere that today's three accepted theories of aging came to be. In the land of the selfish gene, aging could never evolve on its own. Today, the evolutionary mainstream still believes that evolutionary competition is always among individuals, never among groups. If it looks like an adaptation for the sake of the group, it is an illusion, created by selfish genes working on behalf of copies of themselves in siblings and cousins. The new science of aging has to contend with and carefully dismantle the bomb of this dogmatism to make clear the basis of what has long been regarded as an unstoppable, irreversible process. When amateur surrealist and bestselling zoologist Desmond Morris proposed (in *The Naked Ape*) that the breasts of primeval women standing upright in the savanna served to scare away large mammals, he was out on a limb, crafting fairy-tale explanations where none are needed. Ironically, neo-Darwinists seem to have made the opposite mistake: not considering the possibility that aging has been selected for an adaptive advantage. No matter how much evidence points to aging as an evolutionary adaptation, mainstream scientists cannot warm to the idea that the individual sacrifices all for the larger community. Theories of the survival of the fittest individual are simply too deeply embedded in their thinking. It is perhaps ironic that the ideological need to toe the line as part of a group—itself arguably an example of group selection—is at work in their objection.

We must face the paradox that the biologically successful members of our society are to be found principally among its social failures, and equally that classes of persons who are prosperous and socially successful are, on the whole, the biological failures, the unfit of the struggle for existence, doomed more or less speedily, according to their social distinction, to be eradicated from the human stock . . . In societies so constituted, we have evidence of the absolute failure of the economic system to reconcile the practice of individual reproduction with the permanent existence of a population fit, by their mutual services, for existence in society.

—R. A. FISHER

Instant Replay

It is not possible to understand what has happened to Darwin's theory without referring to the cultural context and the sociology of science. Some of the forces that shaped neo-Darwinism were great currents of Western culture, and some were mere happenstance.

The nineteenth century was a time in Europe, England especially, when the landed gentry were losing their grip on legitimacy, and they seized on the philosophy of social Darwinism to justify their position at the top of the heap. How easily they slipped from strong and able individuals prevailing in the struggle for existence to rich and powerful individuals prevailing in the struggle for money and position!

The personal strengths and prejudices of one genius, R. A. Fisher, also played an inordinate role in shaping the evolutionary theory we know today. Much of Fisher's passion for the subject grew from his fears that his kind of intelligence was in danger of being driven to extinction by the rampant reproduction of the masses.

A lesser force that had an equally profound effect on evolutionary theory was that computers had not yet been invented, and a theory based on one gene at a time is one that lent itself readily to equations that could be solved by hand. The complex interaction among genes and the interplay between ecology and evolution are ideas that present-day theorists can model in a computer simulation, but they would not yield to the mathematical and statistical methods known to Fisher.

Through the twentieth century, Fisher's theory was indeed quite successful in explaining laboratory results in artificial selection. This led to the broad perception that there was a predictive, mathematical theory of evolution that had been thoroughly tested against reality and was unassailable. This, however, was an illusion. The lab experiments were designed around the theory. They studied one trait at a time, and selection for each trait was arranged within a static population in a constant environment, as assumed in the theory. That theory works well under these conditions but tells us nothing about whether such conditions prevail in the natural world. It was just circular reasoning.

Then there was the group selection debate of 1966–75, which, in retrospect, was entirely too abstract and resolved in a dangerous victory of theory over observation. The outcome of that debate hinged on both the intelligence and charisma of the personalities arguing for the purity of individual selection and also on the willingness of thousands of field scientists to accept the claims of mathematicians that their equations embodied a pure truth that mere observation of nature could not contravene.

Finally, there is the herd mentality, which has no rightful place in science but which has crept into the scientific community through human weakness. The power structures of funding and publishing bureaucracies serve to amplify these biases.

The good news is that many scientists now realize that evolutionary ecology holds many more surprises than were ever dreamed of within Fisher's philosophy of neo-Darwinism, Horatio! The revolution is already under way.

FOUR

Theories of Aging and Aging of Theories

I think that God, in creating man, somewhat overestimated his ability.

—Oscar Wilde

Aging Theories Confined to Darwin's Straitjacket

All the best science headlines today are about breakthroughs in biochemistry. But just fifty years ago was the golden age of physics, with space exploration and nuclear power captivating our imaginations. It was this zeitgeist that gave rise to a love affair between evolution and mathematics. Biologists came in from the fields and forests and sat at their desks with pads of yellow paper. Evolution was in the thrall of R. A. Fisher, looking for the equations that govern life. From the union of biology and mathematics, the evolutionary theory of aging was born.

In this new way of thinking, natural selection cares only about reproduction, fast and copious. Aging can't help with that. In fact, it's a hindrance—aging detracts from individual fitness. So if evolutionary theory is to account for aging, the possibilities are limited. The three theories that came to light in the ensuing decades may be the only logical possibilities.

1. Aging is beyond the reach of natural selection. No one in nature gets old enough for aging to matter at all. The body falls apart. That's what happens when natural selection isn't looking out for you.
2. The genes that cause aging and the genes that increase fertility are actually the same genes, so that evolution has had to accept aging as a price paid for increased fertility.
3. The body doesn't have enough energy to do everything well, so it is forced to skimp on something. Aging is what results when the body compromises the infrastructure budget in order to boost this quarter's bottom line (reproduction here and now).

Over the past fifty years, all three of these theories have come to be accepted, and most scientists in the field see no inherent conflict among them. In fact, the three together have survived so many empirical contradictions largely because weaknesses of each are reported as if they were validations of the others! But in fact, each of the three theories faces experimental contradictions of its core premise.

August Weismann

In all his writings, Darwin didn't say much about aging. Maybe he hadn't absorbed the message of chapter 2, that thermodynamics wasn't about to explain the issue and it must fall on evolutionary thinking to understand aging. But more likely, he perceived that aging evolved quite generally as part of life's plan, and yet it seemed on its face that aging could only detract from fitness. Thus aging confronted his theory with a core paradox for which he had no ready resolution. Perhaps he didn't think about it at all: Darwin had his hands full with other surprising and fascinating features of the biosphere that not only supported but enriched the theory of evolution, and these probably consumed his attention, time, and study.

Thirty years passed before the first ideas about aging and evolution were seen in print. They were the ideas of August Weismann (1834–1914), an analytic thinker and famous German biologist who in some ways was Darwin's first successor. Weismann's theory is widely quoted today as "making room for the young." Death of each generation assures the opportunity for change and flexibility, facilitating evolution. This idea is rich with implications to which we shall return in chapter 10, but it was not Weismann's. What Weismann wrote was that accidents happen and the body becomes damaged over time. Aging is a kind of garbage disposal service—nature's way to remove worn and damaged individuals so they don't cramp the niche or crowd out their fresh, newly minted offspring. Aging rotates the stock and clears the shelves of stale products.

This idea that aging evolved to eliminate damaged individuals from the population doesn't really make sense. Weismann realized as much himself, so he never developed his theory of aging and distanced himself from it later in life. Peter Medawar said it squarely in 1957: "Weismann canters twice round the perimeter of a vicious circle. By assuming that the elders of his race are decrepit and worn out, he assumes all but a fraction of what he has set himself to prove."

Ernst Mayr identified Weismann as the second-most important nineteenth-century biologist after Darwin. He thought deeply about the cellular workings of life and was the first to realize the importance of the separation between the germ line and the soma. The genes are in the germ line, and they may go on and on with an immortal legacy; the "soma" is the body, and it serves a temporary purpose to safeguard the genes and gather resources needed to reproduce. Thus Weismann contributed the "germ" of such twentieth-century ideas as neo-Darwinism and the Disposable Soma Theory, which we will encounter toward the middle of this chapter.

Weismann is also remembered for having purged Darwin's theory of its Lamarckian vestiges. We now think of mutations as being blind and random, but Darwin hedged his bets on this issue and supposed that the experience of this generation might inform mutations in the next. In a crude test of this idea, Weismann cut off the tails of twenty generations of mice and observed that each succeeding generation was born with tails just as long as the first.

In the twenty-first century, we have seen a recognition of Lamarckian inheritance in the context of epigenetics. Maybe Darwin

was right to play both sides of this issue. This is not the only instance in which Darwin wrote with insights and intuitions that were far ahead of his time. (He also made mistakes.)

Sir Peter Medawar

The brilliant Nobelist Peter Medawar (1915–1987), although he warned he was just an immunologist "dabbling" in theories of aging, is acknowledged as the father of modern evolutionary theories of aging. For half a century after Weismann disavowed his own ideas, aging was indeed a mystery, an "unsolved problem of biology." In 1951, Medawar took this as title for his inaugural lectures at University College, London, and a book that offered a new insight. His idea was the "declining force of natural selection," and it formed the basis of theories on the subject for the ensuing half century. There was to be room for three branches of theory, with many twigs and sprouts to explain the details; but all grew from the taproot of Medawar's insight. Here is the essence of Medawar's "declining force":

Suppose we assume that one set of genes controls your fitness when you are young and another set controls your fitness when you are old. If natural selection acts separately on the "young genes" and the "old genes," then selection will be strong on the "young genes" and not so strong on the "old genes." The first reason for this is that even in the absence of aging, some animals are always dying of disease, predators, and accidents. Those that reproduce early get to pass on their genes; those that reproduce late may or may not get the chance. Natural selection is strongly motivated at early ages, because the individual's entire genetic legacy is at stake. But later in life, the individual may already have succumbed to the slings and arrows of outrageous fortune. There is a second advantage to fast reproduction, even if survival were guaranteed: a shorter generation time means that grandchildren and great grandchildren will appear earlier. Assuming that the descendants all reproduce more rapidly, they will expand to take over the population because their exponential rate of growth outpaces competitors that produce the same number of offspring but spread over a longer time.

Does this explain why aging evolved? Perhaps it is the beginning of an understanding, if we believe the premise that there are different genes active at different times of life, so that natural selection can act differently on "young genes" and "old genes." Medawar, in his monograph, was actually quite modest in his claims for the idea and respectful of the mystery. Theory in the biosciences was still regarded at that time with requisite skepticism. "My proposals can hardly be said to add up to a self-sufficient theory . . . [Weismann's theory stirred] up his successors to think of a more polished and cogent explanation. Not much more than this can be said of any biological theory of comparable pretensions, and I shall count myself lucky if I hear an equally sympathetic criticism of my own."

If Medawar was really trying to be modest for concern over scientific accuracy, it backfired on him. In the years that followed, Medawar's insight was gradually elevated to be regarded as a solution to the "unsolved problem," perhaps because it was the only game in town. His monograph is credited as the foundation of the three modern theories for the evolution of aging: Mutation Accumulation, Antagonistic Pleiotropy, and Disposable Soma. Imposing names, yes, but simple ideas. Though each of these theories was not laid out in name by Medawar, he sowed the seeds. The fully developed theories came out at a leisurely pace, one per decade. Each theory is a mouthful of words, a headful of ideas. Each too, as we'll see, was a sound and worthy hypothesis when first proposed—but later ran afoul of experimental findings.

Theory #1: "Mutation Accumulation"

"Mutation Accumulation"—when you hear this name, you may find it natural to assume that it describes the mutations that accumulate in your cells during the course of a single lifetime until they lead eventually to loss of function, aging, and cancer. This would be akin to the Free Radical Theory dispatched in chapter 1. Although a perfectly understandable interpretation, that's not what this theory is about. Mutation accumulation is, rather, about the *generational* accumulation of mutations—building up to the point where they cause aging but are not, per Medawar, weeded out by natural selection.

Natural selection pays a tax, a "cost of doing business," where her business is trying out mutations, sifting through them, sorting the helpful from the junk, and keeping those few that offer benefits in the form of improved survival or increased reproduction ("fitness"). The great majority of mutations are deleterious, but it's evolution's job to try them, anyway, because that's the only way to find the good ones. So mutations, good and bad, are tested over evolutionary time, and those that don't produce enough progeny to compete with the "better" genes are slowly discarded.

But this means that at any given time, there is a set of new, detrimental mutations that nature has not yet finished eliminating. Indeed they have nothing to offer and only drag down the individual in which they appear, but unless their harm is severe, such genes may linger a long time before they disappear from the population. By the time they do, of course, additional bad mutations will have appeared. Thus the population is always carrying the load of mutations not yet eliminated, and this is called *genetic load.* Genetic load is part of the price of progress, the natural fitness cost of a system that is constantly in flux, seeking ever better solutions.

It was Medawar's insight (made explicit in a later article by E. B. Edney and Robert Gill from 1968) that genes might work on an internal clock in such a way that different genes act on cue at different developmental stages. If this is so, then genetic load would be expected to be higher for genes that act only at later ages. In fact, Medawar realized, many animals in zoos would attain ages almost never realized in the wild, and for such old animals, the selection pressure to eliminate harmful mutations would be just about nil.

So here was an attractive hypothesis for the evolutionary provenance of aging, a solution to this unsolved problem of biology. Mutations are happening all the time. Most mutations are bad. The ones that are very, very bad get eliminated right away, as they make either life or reproduction quite difficult. But other bad mutations stick around for a while before they are eliminated. Bad genes may last a long time if they only affect fitness late in life, after most reproduction has been completed and after most of the population has been lost to predators, disease, or starvation. They will have been passed along as handily as the good genes before their disadvantage has time to play out. Bad genes, if they act only late in life, will accumulate over evolutionary time, causing a multitude of diverse problems that become ever severer with age. This is the Mutation Accumulation Theory.

A possible example often cited today (though none of this was known to Medawar in 1951) is in the genes for dementia and high blood cholesterol. Apolipoprotein E (or ApoE) is an enzyme produced by our bodies for the purpose of breaking down old fat molecules that are ready to be replaced. ApoE comes in three varieties—ε2, ε3, and ε4—depending on details of the DNA that carries the code; in other words, there is a gene for ApoE, and three common alleles (versions) of this gene occur in populations around the world. ε4 raises blood cholesterol and elevates the risk of dementia and heart disease. It's bad for you. ε2 lowers blood cholesterol and depresses the risk of dementia and heart disease. It's good for you. Why, then, is ε4 more common than ε2? Perhaps it's because dementia and heart disease don't affect most people until their prime reproductive years are past, and by then, natural selection "doesn't care so much." So selection against ε4 is rather slow, and evolution just hasn't completed the job of eliminating it. ε4 may be an example of the Mutation Accumulation Theory in action.

Problems with the Mutation Accumulation Theory

There is an extreme version of Medawar's hypothesis, which holds that aging in the wild does not exist at all. It is only in protected environments like zoos that aging can be observed. As it turns out, this is quite wrong. Medawar could not have known this at the time, but in fact, many animals in the wild live long enough that aging factors into their demise. The central premise of his theory—that aging is invisible to evolution—was disproved in field studies more than thirty years after Medawar's book.

Of course, animals in the wild don't become ever frailer and then finally keel over from old age. But that's not the right criterion. It's a highly competitive world out there, and aging can kill earlier and more subtly. A five-year-old gazelle runs a little slower than a four-year-old and finds itself at the back of the herd with a lion in chase. Mice are often strong enough to survive their first winter but not their second. Older fish may not survive a fungal infestation because their immune systems are not as strong as those of younger fish in the same school.

Field studies conducted in the 1980s and '90s looked for remains of a particular kind of animal or bird, assessed the age from the bones, and collected statistics to compute the probability of dying at various ages. If aging were not a factor at all (as Medawar supposed), the probability of dying would be the same for all ages that could be observed. In other words, Medawar's prediction would be that you would rarely or never find in the wild an animal old enough that age was a factor in its death.

That's very much not the case. Almost always, older animals die at a greater rate than younger (mature) animals. The right question to ask: What percentage of all the deaths in the wild would not have occurred but for senescence of the animal as a factor? Answers range from a low of about 10 percent for rabbits and squirrels to 60 percent or more for some alpine and arctic species.

In one heroic study, a young Canadian researcher named Russell Bonduriansky earned his Ph.D. by going out into the forest and labeling individual antler flies, following them around, and continuing to track them until they died. He found that about 28 percent of deaths could be attributed to aging.

Whether we take the answer to be 10 percent or 28 percent or 60 percent doesn't really matter, because even 10 percent is far from being invisible to natural selection. Think of the tiny differences in strength or acuity of hearing or smell that have been honed so effectively by natural selection. Even a 1 percent difference in fitness matters to nature. Aging takes far too large a toll for mutations to be able to stick around in the face of natural selection.

There's another reason, too, why the Mutation Accumulation (MA) Theory cannot be right, and again it is derived from studies that took place after Medawar was dead. In the 1990s, researchers discovered that many of the genes for aging come in families. Aging genes were discovered in yeast cells and worms, and surprisingly, they were *homologues*. This means that they were not identical but close enough that they must have derived from the same source, from a common ancestor. Yes, you and I and the yeast cells and worms and flies all had a common great-great-grandmother less than a billion years ago. Genes that regulate aging are closely related, even among species that are distantly related. And, amazingly, insects and birds and mammals like you and me have our own versions of these same genes. That means that these genes are very old indeed, as they must have come from our last common ancestor, that great-great-grandmother of our deep evolutionary past.

This doesn't square at all with the MA Theory. Remember—the whole point of mutation accumulation is that aging is caused by mutations of recent origin, so recent that natural selection has not yet had a chance to weed them out. From the MA Theory, one would expect that flies age in a very different manner from the way people age, because the random mutations in humans have nothing to do with the random mutations in flies. The homology of aging genes across such different species can only mean that aging has been around a long, long time, has been subject to natural selection, and has *not* been weeded out. This conclusion is not consistent with the MA Theory.

Theory #2: "Antagonistic Pleiotropy"

The second theory to grow from the seed of Medawar's idea is called "Antagonistic Pleiotropy" (AP). It says that the same genes that enhance fertility early in life cause aging later in life. The theory doesn't require all fertility genes to incur an aging cost. But it does require that all aging genes enhance fertility or provide a comparably strong advantage. Now that aging genes have been identified, this does not seem to be the case. A further problem with the theory is that it assumes evolution's hands have been tied. The theory requires that nature has been unable to separate fertility from a kind of superheated, corrosive body chemistry. But experience tells us that biology usually has little difficulty in separately optimizing important functions, even when they are much more closely related than reproduction and aging. There is no law of nature that says owls can't have both sensitive ears and acute night vision, that they must choose one or the other. Why must all living beings have to choose between fertility and longevity? The most direct experimental evidence against this theory is that when fruit flies are bred for longevity, their *fertility goes up*. Hence, whatever genes are causing aging, they can't be enhancing fertility.

"Antagonistic Pleiotropy." Let's deconstruct the term: "Pleiotropy" describes the situation where a single gene has two or more actions in an organism. There is a special word for this because, within the neo-Darwinian framework, it is a case requiring special treatment. Classical population genetics analyzes the effect of fitness from each gene, and multiple, separate effects must be treated as an exception. However, in the real world, geneticists find pleiotropy wherever they look for it. Pleiotropy, indeed, seems to be the rule, and it is tightly targeted genes that are the exception. (There's no word for "one gene, one effect," but perhaps there should be.)

Antagonistic pleiotropy means that one gene carries both a benefit and a cost. The phrase was introduced by Michael Rose, but the idea traces to George Williams, the young theorist who rose to prominence by criticizing the group selection logic of V. C. Wynne-Edwards. Williams's insight was that if there were such genes, where the benefit comes early in life and the cost comes late, then this would handily explain aging. This idea forms the basis of what has become the most popular and best-accepted theory of aging over fully half a century.

Williams was openly skeptical of Medawar's conjecture that no one in the wild died of old age and that aging was invisible to natural selection. Well before there was any data on this question, he (correctly) intuited that the early stages of aging would have a deep effect on an organism's ability to survive in a highly competitive environment. He thought it unlikely that aging could have escaped evolution's filter and felt a more active explanation was called for.

Genes and Timing

The definition of a "gene" has sharpened since Williams wrote up his theory in 1957. Speculating in the absence of our modern knowledge of genetics, Medawar was just before, and Williams just after Watson and Crick unveiled the double helix structure of DNA. The word "gene," now understood to be a stretch of DNA that is transcribed into a protein, was less specific then. For Williams, a gene was the smallest unit contributing to a heritable trait, what Mendel had called a "factor."

We now know that genes are constantly being turned on and off in response to the environment and the state of the body. This is the science of "epigenetics," the winding and unwinding of spools of DNA, the decorations alongside the DNA molecule that determine which genes are active at any given time. Age is one of many ingredients that affect which genes are expressed. In fact, only about 3 percent of our DNA is genes. Most of the rest forms a vast web of signals and targets—"promoter regions"—that determine when and where each gene is turned on.

The idea that genes are turned on and off at different stages of life as part of a highly evolved program may have been unimaginable to Williams. But this is now common knowledge. We now have gene maps that identify which genes act at which stages of life. With the knowledge and concepts available in 1957, Williams had imagined that if a gene were helpful in some places at some times, the body might be stuck with it at other times and places where it wasn't so useful. This was antagonistic pleiotropy. But current knowledge of the intricate dynamics of gene regulation makes this much less plausible.

As a good scientist, Williams listed predictions of his theory—experimental tests that might be done. He was courageous enough to bet his theory on these predictions. Curiously, he didn't predict explicitly that many genes would be identified that have the requisite property (conferring benefit early in life; exacting a cost later on). There was no such thing as DNA sequencing in 1957, and the prospect of cataloging genes must have seemed remote. Another reason he did not make this prediction is that he supposed, on theoretical grounds, that aging should be the consequence of many genes, each of which separately has only a small effect.

Williams stated quite directly that the primary motive for his theory and also the best arguments for its acceptance were based on theory rather than actual identification of pleiotropic genes. "[T]here seems to be little necessity for documenting the existence of the necessary genes. Pleiotropy in some form is universally recognized, and no one has ever suggested that all the effects of a gene need be equally beneficial or harmful, or that they must all be manifest at the same time."

Looking back at his seminal paper from a distance of fifty-plus years, I find it strange that Williams didn't seek to ground the theory more in experiment. Why didn't he consider it important to identify actual genes that were good candidates for pleiotropy, genes that had both a demonstrable benefit in youth and also a clear association with some change that we commonly identify as aging? One hint concerning Williams's thinking is this: "Senescence should always be a generalized deterioration, and never due largely to changes in a single system." Williams expected that the number of multiple-acting genes must be very large, since natural selection would seize every available opportunity to accelerate reproduction, and there must be a large number of such opportunities accompanied by a broad variety of late-life costs. Williams explicitly rejected Medawar's hypothesis that just a few mechanisms would be discovered that underlay all the various phenomena we identify as aging: "Any such small number of primary physiological factors is a logical impossibility if the assumptions made in the present study are valid." So Williams did not propose to look for fertility/aging genes because he expected that they would be extraordinarily hard to spot, each one by itself having only an insignificant effect.

Four decades later, he was still very actively engaged in the field when the discovery of aging genes took off. Since 1990, many single genes have been discovered that have powerful effects on the life spans of lab animals. It is common to find gene variants in lab worms that can increase life span by 50 percent or even 100 percent, all from a single gene. The life-extension record for a single gene in worms is over 1,000 percent.

One prediction that Williams did not make (though it follows directly from his theory) is that fertility ought to rise at the end of life. If there are hormones that cause long-term damage to the body but increase fertility immediately, then the body should use them with restraint when it has a long life expectancy ahead; but when the end is approaching in any case, that's the time to bet the farm on propagating like all get-out. Most likely, Williams realized this was a consequence of his theory but knew it wasn't true, and he thought he would have enough trouble selling his theory without pointing out its deficiencies in the very first paper. In recent years, many animals have been discovered to lose their fertility entirely at the end of their life span, as we saw in the semelparity ("Instant Aging, Sudden Death") section of chapter 2.

Problems with Antagonistic Pleiotropy

One premise of the AP Theory, that there exist genes that are beneficial in youth but lead to cancer or heart disease later on, is perfectly true. Many examples of pleiotropy around life span can be found. Still, the AP Theory is not a plausible explanation for the evolution of aging for two reasons: First, there are many genes that promote aging that *do not* have pleiotropic benefits. Second, the theory only works if it is impossible to separate the benefits from the hazards of these genes; but, in fact, it happens all the time that genes are turned on when needed and off when they are not needed.

A striking thing about aging genes is that they are usually *dialed up* late in life, giving every appearance of a suicide program. For example, consider what we know about fertility and aging in women. When women pass menopause, their female hormones are turned way down, but their risk of cancer and dementia rises at the same time. Two hormones that are *not* turned down are GH and FSH, *gonadotropin* and *follicle-stimulating hormone.* Although they have no use after women are done menstruating, these hormones, after menopause, are turned up so high that they increase the risk of Alzheimer's disease and osteoporosis in women, far above the risk in men.

Multipurpose genes exist, to be sure, but they seem more a tool than an inevitable limitation natural selection has been forced to work with. Nature frequently recycles her inventions, finding ingenious new uses for existing "technology." The logic of AP Theory, however, requires that pleiotropy must be unavoidable, that the price of the ticket for the train ride of fertility is nothing less than an early death. The inconsistent pleiotropy that has been observed should not be interpreted as partial support for the theory. Rather, every gene found that shortens life span without conferring any benefit should be considered a strike against Williams's theory.

Nonetheless, Williams was, in my opinion, a good scientist, and he went out on a limb to make specific predictions from his theory. Of course, some of those "predictions" were actually gross observations about the phenomenology of aging, things that had been known to be true for a long time. For example, he "predicted" that animals that take longer to develop and mature should have longer life spans. (This is true in general, but there are exceptions: remember the cicada that spends seventeen years maturing underground before living out its adult life in a single day.) But in his very first prediction, that aging should be present in all large organisms and absent in asexually reproducing unicells, Williams misfires. In making prediction #1 (see list just below), Williams thought he was making a safe bet, predicting that there should be no aging in one-celled protists. After all, there is no difference between "parent" and "child" when a single cell divides. But Williams also stuck his neck out to make some real predictions, some of which could not be tested for decades afterward.

Predictions made in the paper proposing AP:

1. Aging should be universal in macroscopic organisms, absent in one-celled creatures and cloners.
2. Low adult accidental death rates should be associated with long life spans.
3. Animals in which adults increase in fecundity should age more slowly.
4. Where there are sex differences in death rates, higher mortality in one sex should be associated with shorter life span.
5. When an animal's systems fail with age, everything should fail at once; no single gene or even a single system should be able to prolong life span.
6. After fertility ends, with no evolutionary pressure to keep them alive, animals in nature ought to die promptly.
7. The first signs of aging should appear at the time of sexual maturity.
8. In lab experiments, artificial selection for increased longevity should result in decreased vigor and curtailed fertility in youth.

So, how have these predictions fared over the years? By and large, miserably, not to mince words. But this embarrassment has created opportunities for industrious theorists to extend or modify the original theory time and time again. Tests of Williams's first seven "neo-Darwinist" predictions are summarized in the box directly below; the eighth we'll look at separately.

Seven Predictions of George Williams (1957), and How They Fared

1. Aging should be universal in higher organisms, absent in one-celled creatures and cloners.

Wrong. **We know from chapter 2 that there are some animals and many plants that don't age. Williams probably thought it was a safe**

bet, based on what was known in 1957, that single-celled organisms don't age. But as we'll see in the next two chapters, there are two modes of aging in single-celled organisms that cannot be explained by AP Theory.

Even some bacteria incorporate aging into their life cycles. None of this can be explained by Williams's theory or by the other standard and accepted theories of aging. Are we to think that aging in microorganisms evolved on a completely different basis and for a different reason than aging in multicelled life?

2. Low adult accidental death rates should be associated with long life spans.

Perhaps. There have been some experiments that confirm this expectation and others that confound it. When opossums were transported to an island where there were no predators, they evolved longer life spans. But in Trinidad river pools where guppies are preyed upon by bigger fish, the animals actually evolve longer life spans than in nearby pools where there are no predators.

3. Animals in which adults increase in fecundity should age more slowly.

Correct. This prediction has been borne out. But unanticipated by Williams, some animals continue to grow more fertile over their whole lives, and they "age in reverse," in the sense that their chance of dying goes down, down, down from one year to the next.

4. Where there are sex differences in death rates: higher mortality in one sex should be associated with shorter life span.

Maybe. This prediction has yet to be tested systematically.

5. When an animal's systems fail with age, everything should fail at once; no single gene or even a single system should be able to prolong life span.

Wrong. One of the great surprises in the field of aging genetics has been how easy it is to find single genes that can be modified to extend life span. Williams conceived of pleiotropy as thousands of tiny bargains—a Faustian dynamic, if you will—in which the integrity of one bodily system after another was traded away for a little increment in fertility. Contrary to the expectations of Williams and everyone else, aging seems to be controlled in a big way by a handful of genes with large effect.

This, in fact, is a major clue about the nature of aging. Genes are organized in hierarchies, with a few master genes controlling the broad currents of growth and development. These same master genes are also connected to aging. This suggests that aging evolved as part of the life cycle, like development, growth, and sexual maturity.

6. After fertility ends, with no evolutionary pressure to keep them alive, animals in nature ought to die promptly.

Wrong. (Some examples were raised at the end of the previous chapter.) We are all aware of the counterexample of menopause in women. Anthropologists like to explain this in terms of older women's devotion to the upbringing of their grandchildren. But surprise—life after the end of reproduction turns out to be ubiquitous in the biosphere, including in such animals as worms and even yeast cells that don't give a shilling or a whit about their grandchildren. Whales, otters, opossums, elephants, guppies, quail, parakeets, and mice have all been reported to live on after they lose their fertility.

7. The first signs of aging appear at the time of sexual maturity.

Sometimes. It's true that no animals show signs of aging before sexual maturity,* but many only begin to age well after they have begun to reproduce.

* Except perhaps in the genetic malady *progeria*, where some traits (wrinkles, frailty) appear in young children. First described in 1886, progeria, which is the result of new mutations, as it kills before it is passed on, affects about one of eight million live births.

Williams's Prediction #8

8. In lab experiments, artificial selection for increased longevity should result in decreased vigor and curtailed fertility in youth.

There have been two main ways to test AP. Fruit flies have been the experimental animal of choice for both. First, breeding experiments have been done to see if fertility and longevity are related. The expectation was that, bred for great longevity, fruit flies would suffer catastrophic fertility loss. The second means was to probe natural variation. Do the individual flies that live longest tend to lay more eggs or fewer?

Looking for Multi-Acting Genes in Laboratory Breeding Experiments

In the first category, the marathon experiment has been carried out by Michael Rose's lab at the University of California–Irvine. In the 1970s, Rose was a star student of a great British theorist of mathematical evolution, Brian Charlesworth. Rose, too, is a mathematical theorist at heart, but he also realized what the field needed was an experimental basis for the heady science. So he set about in 1981 to create just the experiment described by Williams that would validate the central premise of the theory of pleiotropy.

As a young professor in Nova Scotia (and later Irvine, California), Rose began breeding fruit flies for longevity. His method involved no hi-tech genetic engineering but a straightforward application of the way that plants and domestic animals have been bred for hundreds of years. He maintained his flies in jars until 90 percent of them had died and then collected eggs from the remaining flies. The next generation of flies would live a little longer, and he repeated the same procedure with them as longevity progressed with each generation.

The beauty of this experiment is that it could be done without knowing what genes affected aging. In order for the experiment to work, all that is required is that there must be some genes that affect life span, and there must be some variation among the flies in the distribution of these genes. Internally, what must be happening is that various genes for longevity become more and more concentrated in each successive generation.

When Rose began his experiment, each generation of flies lived for two weeks. But the life span continued to advance, and as I write in 2015, his experiment continues, but with flies that live more than sixteen weeks.

The very fact that he could breed very long-lived flies was surprising. There were certainly no flies that lived anywhere close to sixteen weeks in the population that Rose began with in 1981. We imagine that what is happening is that many kinds of longevity genes were spread through the population, so that some flies had longevity gene A and some had longevity gene Z, but none had a full complement of all the longevity genes. Breeding is a (slow and laborious) way to combine these genes and gather them together in a single individual. It is this process that led to longer-lived flies than any in the initial population.

The big objective in Rose's experiment—the subject of his work and the reason that he embarked on the experiment—was an attempt to detect genes that were good for fertility but caused aging later on, a.k.a, pleiotropy. Rose fully expected that over the years as his inbred flies lived longer and longer, they would grow less and less able to reproduce. He even imagined that ultimately, the limit on the experiment would be reached when he was able to breed flies that had superlong life spans but that laid no eggs at all from which he could breed a next generation.

Two years into the experiment, there was a dip in fertility, and Rose rushed to publish a paper in which he announced that the signature of antagonistic pleiotropy had been detected. "The scientific significance of these conclusions is that senescence in this population of *Drosophila melanogaster* appears to be due to antagonistic pleiotropy, such that genes which postpone senescence appear to depress early fitness components. Put another way, these results corroborate the hypotheses of a cost of reproduction (Williams 1957, 1966), since prolonged life seems to require reduced early reproductive output."

But soon the results began to turn around. After two more years, the finding was undeniable: the long-lived flies were laying more eggs than the control flies (that continued to live for only two weeks).

* * *

AP Theory says that the genes for longevity and those that depress fertility are the same—that's why natural selection over the aeons hadn't done the job that Rose's lab had done in a few years. But the experimental result seemed paradoxical: these same genes that increased longevity were also enhancing fertility. How could that be? Where was the pleiotropy? More to the point, why had natural evolution failed to find these same super-fly combinations that Rose bred in such a short time?

Rose the theorist stepped forward with proposed answers to these questions. But he has not faced to this day the fact that the central prediction of George Williams has failed spectacularly or held the theory accountable for its failure. The explanation that Rose posed when he published this result is that while he was breeding for longevity, he was also, incidentally and unintentionally, breeding for fertility, as well. There may have been many infertile flies in that last remaining 10 percent of each generation. But these flies didn't contribute eggs to the next generation; only flies that were both fertile and long lived were selected by his procedure.

This is an unsatisfying answer, however, because it doesn't address the question of how there can be such flies (that are both more fertile and longer lived) or why it is that nature's own evolutionary process has not identified them. It does not address the fact that not only do Rose's superflies lay more eggs in old age but they are more fertile in their youth than the flies in the control bottles that live only two weeks. The flies that Rose has bred not only live for sixteen weeks compared to two weeks for the controls but on every day of their lives, they lay (on average) more eggs than the control flies lay on the same day.

Looking for Multi-Acting Genes in Wild Populations

The second mode of testing AP has been realized by many scientists who looked for a relationship between fertility and longevity in animal populations. Instead of taking many generations to breed animals for large differences in longevity, they have examined the natural variation within populations. Some animals live longer than others; some animals are more fertile than others. AP Theory predicts that these two traits should be inversely related. When you find an animal that lives extra long, it should tend to leave fewer offspring. Conversely, the animals that leave the most offspring should tend to have shorter life spans. When statistics are compared, this effect would appear as a negative correlation between life span and fertility.

The best design was by four scientists (the Minnesota Four) at the University of Minnesota in 1995–96. Their experiment used one hundred bottles of inbred fruit flies, where the flies were all genetically identical within each bottle but different from one bottle to the next. They tracked both fertility and death rates day by day. Their main finding was that there was a *positive* correlation between longevity and fertility—the opposite of the theoretical prediction.

Here's a little irony of aging research that you are now in a position to appreciate. The two most prominent dueling theories, as of 1995, were those of mutation accumulation and antagonistic pleiotropy. The Minnesota Four tested both of these when they analyzed their data from the hundred jars of fruit flies. One paper they published focused on genetic diversity to test the MA Theory. It was written up with Daniel Promislow as the first author, and he concluded that the test didn't come off so well for MA, so this should be interpreted as supporting the AP Theory. The second paper focused on correlations between fertility and longevity, to test for AP. It was written up with Marc Tatar as the first author, and he concluded that the test didn't come off so well for AP, so MA was probably the right theory.

Putting the two experiments together, what we really have is devastating evidence against both the AP Theory and the MA Theory.

Smart Theorists, Good Theories—But the Evidence Didn't Agree with Them

Neither Williams nor Medawar had available our modern perspective on gene regulation—that expression of genes is controlled by signaling networks of exquisite complexity and that the time and place of gene expression is adapted under the control of natural selection, just as the forms of the genes themselves are selected. Both imagined false constraints on evolution—ways in which evolution was stuck with fundamentally limited genetic machinery. Medawar imagined that each gene came with a time stamp that determined when in life it would be turned on. (Those genes that had late time stamps could slip by evolution even if they did great damage.) Williams imagined that genes were either always on or always off. (Thus, a gene that offered benefits early in life could not be turned off if it caused harm to its bearer late in life.)

We should lose no respect for Medawar or Williams because they put forward plausible hypotheses based on the knowledge of their day. But we do need to reevaluate the viability of their theories from a modern perspective. In light of all we know today, neither AP nor MA is tenable.

Theory #3: "Disposable Soma"

The third theory can be considered an application of AP to food energetics. Disposable Soma (DS) Theory, also about forced compromises, differs as to when the compromise takes place. AP imagines compromises built into the genome over evolutionary time. DS suggests compromises, forced by metabolic limitations, operational through the individual's lifetime.

Though the theory goes by the intriguing name Disposable Soma, it is really about the body's energy budget. It says that there isn't enough food energy to do everything the body needs to do—to forage and to compete for mates and to run a metabolism and to reproduce and simultaneously to repair the cells that get damaged. The damage is not perfectly repaired because the body can't spare the energy, and it can't spare the energy because of this grand compromise among competing demands.

The theory sounds so sensible and plausible, we may think intuitively it must embody some basic truth. But there is one clear reason it cannot be so. We know that animals that eat less actually live longer. If aging were caused by a shortage of food energy, then more food energy would support the body to repair and preserve itself. The DS Theory predicts that overfed animals ought to live longer than underfed animals, but exactly the opposite is true.

Calories are the currency of the body. Running from predators, fueling the brain, and hunting for more calories all require calories. In the winter, food may be burned just to keep warm. Repair and maintenance of the soma (body) is a task that must compete with these others for the body's allocation of food energy. But the payoff is in reproduction. Reproduction trumps all other uses of caloric energy. The other uses are all indirect applications, investments that may be rewarded with future reproductive success.

DS is the brainchild of English biologist Tom Kirkwood of Newcastle University. Kirkwood realized that every trade-off requires compromise, and compromise means that neither one side nor the other gets everything that it wants. As long as there are competing energy demands, there can never be enough energy to do an ideal job. This appears to be a proof (another proof!) that aging is inevitable:

- Maintaining the body and keeping it in good repair requires energy.
- But other aspects of metabolism, especially reproduction, also place demands on the body's energy supply.
- We expect that natural selection has optimized the allocation of scarce energy for all these tasks, so no one task has received all the energy that it needs.

DS is like the Faustian trade-off of AP—live now, pay later—but it doesn't depend on the way that genes work over time. So it frees AP from the requirement of "one epoch, one gene" that is its conceptual Achilles' heel. DS requires only that our bodies' energy allocation be the result of an evolutionary optimization process. This seems eminently reasonable.

The "proof," however, is another story. The logical fallacy is to imagine that the newly minted adult is in a pristine state that can only go downhill, no matter how much energy is spent to maintain it. But the body never was perfect in the first place, and it requires no perfection to keep it in good repair. The appeal of the proof depends on the idea that keeping the body in exactly the same condition is a kind of ideal, a limiting case that can only be achieved in theory but never in reality. But remember that the body had no trouble building itself from seed, and that process required a great deal more energy than simply maintaining itself after it was built. During that time when growth required an intense energy expenditure, the body had not yet begun to suffer the decline of aging, and in fact it was getting stronger, more fertile, more robust all the while. If building the body strong and fertile was not too difficult or costly, then maintaining it in that state should be far easier.

In Kirkwood's 1977 paper proposing the Disposable Soma Theory, he focused on the integrity of DNA, which is the body's central repository of information. He built his theory on Leslie Orgel's hypothesis (from chapter 1) that errors in copying DNA during an individual lifetime could be a crucial part of the aging process. It was from thinking about information in DNA that Kirkwood got the idea that error accumulation was a one-way street that could drive aging. True in theory. But remember from chapter 1 that DNA errors turned out not to be an important driver of aging, and that DNA integrity is easily maintained over a lifetime without a problem. We now know that chemical damage is indeed a common feature of the aging cell, but it is secondary molecules—proteins, fats, and sugars—not the DNA, that are commonly damaged. The information lost from DNA is so small as to be insignificant. This effectively killed the Orgel Hypothesis, but the Disposable Soma Theory has survived by shape-shifting.

Step back and look at the process of growth and development for a different perspective. It has taken a certain amount of energy to build a body with the strength and endurance and fertility of a twenty-year-old. DS is based on the idea that the body then has two choices: to maintain its pristine state via enormous energy expenditure or to permit slow deterioration, conserving some energy for other uses. But there is a third possibility. The body could grow larger, stronger, more fertile, and more robust by continuing the same process of growth and development. It becomes clear that maintaining the body in its present state is not the ideal of perfection but only a middle ground. The body might grow stronger or allow itself to deteriorate or allocate just enough energy for a steady state. Maintenance of the status quo is not a limiting case or a maximal condition but only an intermediate compromise.

A tree uses some of its energy for present reproduction and may produce hundreds of thousands of seeds in a year. In addition, and at the same time, it allocates enough energy to grow a little larger, a little stronger, and more fertile from one year to the next. Why don't our bodies behave in the same way? Instead of weakening and becoming a little more likely to die from one year to the next, we would be a little stronger, a little more fertile, a little less susceptible to disease, year after year. This is exactly the point made (in the abstract language of mathematics) by Annette Baudisch and James Vaupel in the "sproof" article we described in chapter 2.

The Disposable Soma Theory continues to be popular because it is intuitively satisfying. Researchers assume that it must be true because it is so logical, so easy to understand. In science, hunches based on intuition are a good place to start. But a good scientist must also be willing to abandon his hunches in the face of real-world evidence. In fact, DS starkly contradicts experiment at every turn. It has no evidentiary support, and its predictions baldly fail. The theory predicts that eating more should relieve the body from having to make a difficult choice between fertility now and longevity later. Food should buttress and enhance health and life span, but across the board, animals that are fed less live longer. Since activity consumes energy, animals should pay a price in terms of longevity for physical activity, but animals (and humans) that expend more energy in exercise benefit, both with greater health now and with longer life span. So, too, Disposable Soma Theory predicts that animals that siphon their energy into reproduction must pay a price in decreased longevity. But, in fact, studies of zoo animals show that those that reproduce live just as long as those that have no offspring. And some, like the Blanding's turtle and lobsters, grow ever larger, more fertile, and less likely to die, despite ongoing expenditure of reproductive energy. The theory is appealing but the facts do not add up. Everything but a neon sign in the sky tells us that the need to ration food energy is not the cause of aging.

More Problems with Disposable Soma

Like AP, DS also makes predictions, but, unlike AP, *all* the predictions of DS fail. According to DS, women should live shorter lives than men because they spend far more energy on reproduction. The opposite is true. According to DS, the more children are born to a woman, the shorter should be her life span. But demographers tell us there is a small positive relationship between a woman's fertility and her life span. (In animals, no consistent pattern has been found one way or the other.)

Most egregiously: DS predicts that eating more food should lead to a longer life span, because more total energy means less pressure for the body to skimp on the energy requirements of biochemical repair. But in lab experiments, animals age more rapidly the more they eat, and in fact, conditions near to total starvation lead to a substantial enhancement in life span.

A good way to understand the difference between Disposable Soma and the classic, AP version of trade-off theory is to look at the predictions about fertility and fecundity.

Technical Definitions: Fertility and Fecundity

***Fecundity* is the body's potential for reproduction.**

***Fertility* is actualized reproduction.**

You may say that a thirteen-year-old girl is very fecund, but if she remains a virgin, her fertility is zero.

Classic AP aging theory says the body is "designed" (selected) to be fecund at the expense of longevity. But since the compromise is in the genes, it may not be possible to evade the compromise simply by not having children, since the problem is in the genes and not the behavior. The Disposable Soma Theory, however, sets itself up for a fall: by saying that it is the actual energy spent in reproduction that detracts from longevity, it is far more easily falsified. DS makes a clear prediction that fertility (actually having babies) is itself a cause of aging. Laying eggs, lactating, producing sperm, and engaging in mating contests for males are all included, and these activities ought to decrease longevity.

The same reasoning applies to food. AP makes no clear prediction about behaviors or environmental influences on longevity, because the compromise is already there, built into your genes. But Disposable Soma Theory says that the body suffers aging because it doesn't have enough energy to take care of itself. Eating more means more food energy coming in, and it should solve the problem, eliminate the need for compromise. DS clearly predicts that animals should live longer the more they eat.

Does Having Babies Really Make You Older?

Reproduction requires energy. In the logic of DS, this energy is subtracted from what the body can use for repair and maintenance. So DS predicts that reproduction ought to shorten life expectancy, and Tom Kirkwood has himself devoted considerable energy to trying to prove that it actually does so.

But there's a great deal of evidence from studies not just of people but animals, too, that says if anything, there is a slight *increase* in longevity from having offspring. There is exactly one study that says boldly that having children has shortened women's life spans, and an author of that study is Kirkwood himself. First, let's look at the majority position.

Caleb Finch is a scholar at the University of Southern California known for his encyclopedic summaries of the experimental data on aging. Back in 1990, he wrote a book about aging and genetics that is so big and well researched that it is still the go-to reference today, despite the explosion of research that has taken place since then. Summarizing evidence to that point, Finch said there is a great deal of evidence that animals die in the act of reproducing. (Many women used to die in childbirth, too, right through the nineteenth century.) But there is no evidence that those who survive have foreshortened life spans. Finch interprets this to mean that reproduction entails an immediate risk of death, but for those who do not die in childbirth, that bears no relationship to aging.

It is most convenient to study this question in zoo animals. Often, studying zoo animals is a fallback that offers data that's not what you really want to know, but it's so much more practical and less labor intensive than finding and observing animals in their natural habitats. But in this case, zoo data is not just more convenient, it's more relevant than observations of animals in the wild. This is because animals in zoos have good medical care and die of old age, so we can get a realistic idea of their life spans under standard conditions. Zoos keep good records, and some captive animals are bred, while others are not.

Robert Ricklefs is a biologist at the University of Missouri and author of a prominent textbook on ecology who likes statistics and has written a great deal about aging from the vantage of ecology and demographics. His 2007 study of seventeen mammal and twelve bird species in captivity is the best data we have on the relationship between fertility and aging. "We found no evidence that reproduction early in life influences, either positively or negatively, lifespan in zoo populations . . . Our analyses failed to reveal a significant relationship between number of offspring produced up to a given age and subsequent survival in either birds or mammals."

And what about people? Do women who never have children live longer? Do women with lots of children die earlier? The famous statistician Karl Pearson (his name is still attached to the correlation calculation that is the bread-and-butter of statistics) did the first study, mathematically solid, but not so systematic as would be demanded in an article published today. Selecting birth and death records in Britain and America wherever he could find them, he found a small *positive* relationship between number of children that a woman has and how long she lived. Several other studies early in the twentieth century also found a positive relationship. Since 1990, there have been more and larger tests of this question. Two studies of French Canadians found no relationship. A large-scale contemporary study of Norwegians (2008) found a positive correlation between fertility and longevity. A 2006 study looked at historical records of an American Amish population and found a significant positive connection between number of children and life span.

Thomas Perls of Boston Medical Center studies aging by learning all he can about long-lived humans. He has interviewed hundreds of centenarians and more recently has sequenced their genomes. One of the earliest studies he published from this population was based on a striking observation: his centenarian women were four times more likely to have continued having children into their forties than other women of their generation. Something about bearing a child late in life contributes dramatically to improved chances for longevity.*

The Disposable Soma Theory stakes its credibility on the idea that bearing children should accelerate aging and shorten life expectancy. But there have been many studies either finding the opposite—that bearing children contributes positively to longevity—or finding no relationship at all. Against this background, the originator of the Disposable Soma The-

* Perls did not interpret his finding as a challenge to theory. Perhaps, he said, women who remained fertile into their forties were those endowed with longevity genes. In 2012, I published a paper doing the math to show that this explanation cannot fly. There are just too many women who are fertile in their forties and too few that go on to live to age one hundred.

ory, Tom Kirkwood himself, conducted a study in 1998. Kirkwood found the opposite result from everyone else and announced a triumph for the theory. His was published prominently in *Nature* and led to headlines and press releases in science news outlets everywhere. The Kirkwood study is still cited more frequently than all the other studies combined.

How could this be? Well, Kirkwood used an unusual statistical test. I personally reanalyzed his data using more straightforward methods and got the opposite result from Kirkwood.

The statistical test Kirkwood used (called "Poisson regression") gives unique emphasis to outliers. Out of three thousand women in his database (upper-class British women, going back to the twelfth century), only nine women had fifteen or more children, and of those nine, there were five that lived before 1700 and had short life spans typical of that era. I removed just those five women out of three thousand and then applied the same odd statistical test that Kirkwood used, and without those five women, again the test led to the opposite result.

As the flamboyant Nobel physicist Richard Feynman said in a commencement speech to the graduating class at Caltech in 1974, "The first principle is that you must not fool yourself—and you are the easiest one to fool."

Disposable Soma and Caloric Restriction

In thousands of lab experiments through eighty-five years, cutting back on food has proved to be the most powerful and consistent intervention that we have found for making an animal live longer. This is a deal breaker for the DS Theory, which is built on the idea that the root cause of aging is an insufficiency of food energy.

The connection between less food and longer life predates modern scientific study. Hippocrates hints at it. In fifteenth-century Venice, Luigi Cornaro wrote a volume titled *Discorsi della Vita Sobria* (*Discourses on the Temperate Life*) about his personal experiments with caloric restriction, supplemented by half a liter of wine daily. Cornaro lived to 102. In 1733, Benjamin Franklin (who published a translation of Cornaro's book, with a blurb by George Washington) wrote in *Poor Richard's Almanac*, "To lengthen thy life, lessen thy meals."

The first formal study of caloric restriction was performed by Clive McCay, an American nutritionist, biochemist, and gerontologist at Cornell University. During the Great Depression, there was serious concern about food shortages leading to shorter life spans. McCay's work on food restriction was funded by a private foundation grant. The experiment was designed around the idea of seeing whether he could halt growth in juvenile rats by feeding them just enough to keep them alive but not enough for them to grow. Many a young rat died of starvation before he figured out how much food was just enough to keep them from starving, and he never did succeed in keeping them from growing. But the striking result was evident from the earliest trials. It was clear that the animals that escaped starvation were not living abbreviated lives. They were surviving longer than any rats in the history of science. McCay was surprised, and he repeated his experiment many times before publishing it. The experimental design at which he arrived emphasized good nutrition, with concentrated sources of vitamins, minerals, and protein in a low-calorie chow.

McCay's experiments produced dramatic results, but they were not recognized as being important, and there was little follow-up for almost half a century. The modern rediscovery of caloric restriction (CR) came when Roy Walford served as "house doc" in the Biosphere 2 experiment in Arizona, 1991–93. This was to be a hermetically sealed environment in which a team of bionauts grew all their own food, recycled all their water, and even used photosynthesis to supply their oxygen. But farm productivity was way below expectations, so the crew didn't have enough to eat. Walford noticed that underfeeding the bionauts had dramatic health benefits, though it made them irascible.

Beginning in the 1980s, experiments have been done with yeast cells, worms, fruit flies and other insect species, spiders, crustaceans, fish, various rodents, dogs, horses, and rhesus monkeys. A project at Washington University has followed the health histories of people who practice CR, though mortality comparisons may not be available for a long while. Shorter-lived animals tend to show greater proportional life extension, but health and longevity benefits were discerned even in the monkeys, with a life span of twenty-five years.

* * *

McCay had guessed that a biological clock delayed development if young animals did not receive sufficient food early on, and this same clock might go on to slow aging. McCay's guess about a biological clock turned out to be only partially correct. CR works even if begun in adult animals. Life extension is not as great as when begun earlier, but qualitatively, the effect is the same. His guess about full nutrition turned out to be off the mark, as well. Animals provided with complete nutrition tend to avoid certain diseases, but in subsequent experiments, full nutrition was shown to be a mixed blessing. Life extension has been reported when animals are fed a diet deficient in protein, even if they have plenty of calories. And severe shortage of one particular protein component, an amino acid named *methionine*, can extend life span all by itself.* All of these forms of deprivation extend both mean and maximum life span of a group, so they have been regarded by the life extension community as the "real McCay."

Short-lived and simpler animals tend to show (proportionately) a better response to CR than long-lived animals. Lab worms fully fed live only twenty days, but if starved early in life, they go into a suspended state called *dauer*, which is something between hibernation and a spore. Dauers are extra tough and resistant to heat, cold, dehydration, and other things that normally would kill a lab worm, and they can survive up to four months without food. A dauer is alive just enough to detect food and water in its environment, and when it does, it picks up life and growth from where it left off. Fruit flies may live thirty days with full feeding, but nearly twice as long (fifty days) under CR.

Lab mice normally live two years, and with a severely restricted calorie intake, they can survive almost three. Dogs typically live ten years, and CR offers an extra two years. The longer the life span, the less the proportional gain. Rhesus monkeys in cages have a life span of about twenty-six years, so two American experiments with caloric restriction that were begun in 1990 did not report preliminary results until 2012. By that time, no one really expected to see the huge 80 percent increases typical of fruit flies or even the 40 percent gains that can be found in mice. CR monkeys were clearly healthier, more active, better looking, and disease-free for longer,

* Methionine restriction is not practical for humans. Some people practice protein restriction, but there are downsides, especially if you exercise. My personal suggestions for a longevity diet are in chapter 9.

but life extension was difficult to measure for several reasons. The monkeys were bored and anxious in captivity and tended to become violent, especially those who were deprived of food. The numbers were too small for a good statistical sample. Some reporting in the popular press would lead you to think that the long-term experiments in rhesus monkeys found little or no gain in longevity, but I've viewed the results as positive, all things considered, despite the complications.

In 1978, when Kirkwood came forth with the Disposable Soma Theory, he did not know about caloric restriction experiments. I—and a lot of other people—didn't, either. Although experiments in CR were already more than forty years old, the medicine of aging was still a scientific backwater, and CR was a subfield within a subfield. But since 1996, CR has been at the center of the mainstream in the literature of aging, and the contradiction with the DS Theory is—I know I've been harping on this—a glaring one. It might be an honorable gesture to retract the theory, to say that it was an attractive hypothesis but experiments previously unknown to the author have made the theory untenable. Instead, Kirkwood, working with his Ph.D. student Daryl Shanley, published an article claiming to reconcile the CR data with the Disposable Soma Theory.

Shanley and Kirkwood analyzed the energy budgets of pregnant and lactating mice. Females take in a lot more calories when they are mothering, but after subtracting out the energy used for reproduction and nursing, they found that there was actually less food energy available for repair/maintenance, compared to the same mouse eating less but not reproducing. After taking reproduction into account, less is more. Kirkwood claimed that this resolved the conflict between DS and CR data.

In a rejoinder, I questioned the relevance of their calculation. Comparing mice that reproduced and ate a lot with mice that did not reproduce and ate a little was beside the point. For seventy years, CR experiments had been done explicitly with mice that do not reproduce. Female mice with less food were compared to female mice with more food, and the ones with less food lived a lot longer. This was true even though none of the mice were reproducing. The same thing was done with males, which spend much less energy on reproduction. Males were housed in separate cages—no one to mate with. Some were given more food, and some were given less food, and again the ones with less food lived longer. Kirkwood's comparison missed the mark.

As I write in 2015, the Shanley/Kirkwood article has been cited by 183 articles and continues to be cited each year, usually by scientists who read the conclusions in the abstract and assume that the details in the article support those conclusions. The theory is very appealing, and people would like to believe it.

Salvaging the Disposable Soma Theory?

The biggest weakness of the DS Theory is that its predictions about the body's energy budget are all wrong. Might there be other scarce resources than energy that have forced evolution to forge compromises? Kirkwood proposed the theory in terms of caloric energy and continues to describe it in this way. Other prominent gerontologists have adopted the spirit of Kirkwood's theory, however, applying it to generalized scarce resources without specifying what they might be. The general concept that the body must make compromises seems appealing; the specifics of why that might be and how it might work remain elusive.

Steven N. Austad is a researcher who recently moved to the University of Alabama from the Barshop Institute for Longevity and Aging Studies at the University of Texas Health Science Center. He is a scholar and eloquent advocate for the general idea behind the Disposable Soma Theory. He has coauthored several papers with Kirkwood. While not completely subscribing to any of the three flavors of aging theory, he makes the broadest case possible for trade-offs, filled with wide-ranging detail. His book, *Why We Age*, is accessible to any reader of this book and makes an excellent introduction to traditional theory on the subject.

Austad is the best advocate for the modified Disposable Soma Theory, based not on scarce energy but some other scarce resource. This sounds very attractive, and it avoids the biggest problem with Kirkwood's version. But Kirkwood is right that energy is universally the limiting resource in biology. Plants' growth is proportional to the amount of sun they get. Animals' growth is dependent on the food they consume. No other biological resource is as universally important as energy, and so we know that natural selection has economized and optimized energy usage in a uniquely focused way. But if energy is not the scarce resource that forces the body to compromise and skimp on the infrastructure budget, then what is that resource? Neither Austad nor anyone else is forthcoming with an answer, and until they are, this version is not a theory that can be tested.

Hormesis, or Eustress

Life extension through caloric restriction is truly a striking fact of nature. We may take it for granted that overeating is unhealthy so that we may have never paused to think just how strange a thing it is. If the body is able to keep itself healthy on a meager allowance of food, why would it do less well when given more food to work with?

There's no intrinsic reason why we should expect that carrying around extra weight should be much of a burden, especially if there's extra resource for creating muscle and bone to support it. Elephants live a lot longer than giraffes. And even if there were some metabolic reason why storing so much fat must be intrinsically unhealthy, then why wouldn't the body just discard the extra food energy with the stool or burn it less efficiently? It's strange that the body would allow itself to be damaged so by food.

The mystery deepens when we realize that the relationship between starvation and longevity is just one example of a much more general finding: life spans can be elongated by moderate stress. This is the phenomenon called *hormesis*, or *eustress*, which has been demonstrated in many different contexts; yet it remains controversial because theoretically it is so unexpected. Starving to death constitutes severe stress, but near-starvation leads to a longer life span as well as dramatically lower risk of heart disease, cancer, and diabetes. So from the standpoint of hormesis, the definition of "moderate" stress can extend pretty deep.

Hormesis carries a deep theoretical message. If the body is able to prolong life under stress, despite the burden of the stress, then this can only mean that when the body is not stressed, it is holding something in reserve and not doing its best to prolong life. In this sense, the body is purposefully withholding the repair and maintenance that would help it to live longer.

Another example of hormesis that won't surprise you (except that you never thought of it as hormesis) is exercise. Exercise lowers risk of many (most) diseases, including infectious diseases. In animal studies, physical activity increases the average life span of a cohort. Insurance companies know that sedentary lifestyles are a mortality risk factor and calculate life insurance premiums accordingly, and many enlightened employers will also promote exercise for their employees. So the association of exercise with longer life is certainly familiar, but that doesn't mean that it is expected or even understood by medical science.

Exercise takes a lot of energy. If it's energy that limits the body's repair function (as in the DS Theory), then more exercise means there's less energy available for other jobs. And the energy involved is not trivial. The evidence is that life span goes right on increasing at levels that most of us would hardly consider "moderate" exercise. In fact, elite athletes live longer on average than people who follow current medical recommendations and work out at the gym three times a week. In rodent studies, the mice that live longest of all are simultaneously calorically restricted and on the most intense exercise regimen. These mice voluntarily run two full miles in their treadmills every day.

The fact that exercise extends life span is surprising from other perspectives, as well. If we think of aging as accumulated damage, then we must recognize that exercise adds to the repair burden. Muscles are torn when they are used intensely. This is the stimulus that signals the body to rebuild them larger and stronger. But this rebuilding demands that the body replicate cells, copy DNA, check for errors, and perform all the functions usually associated with biological repair and maintenance. Moreover, exercise generates copious free radicals. Biomolecules are incidentally oxidized (damaged) in the process of rapid respiration and energy generation. This is exactly the sort of damage that is said to accumulate as we age. Somehow, when we exercise, the repair mechanisms become so much more effective that they more than compensate for all the extra damage that is being inflicted so that a net benefit for longevity results. You have to wonder: If these superefficient repair mechanisms are available when the body is stressed by exercise, why don't our bodies deploy them all the time?

Maybe the body isn't trying to live as long as possible. Maybe there is a reason nature prefers to have higher death rates in less challenging times and lower death rates under stress.

Besides exercise and starvation, there are other forms of stress that have the paradoxical effect of prolonging average life spans. Historically, the first to be studied was radiation hormesis. Animals exposed to small quantities of radioactive material, or to small daily doses of x-rays, live longer than animals that have no exposure. Once again, this is completely unexpected. The radiation damages biomolecules, most notably DNA, killing some cells and requiring repair in others. Why should this lead to longer life spans?

Radiation hormesis is an interesting case. In the 1950s, when nuclear power plants were first contemplated on a large scale, government regulators were charged with setting safety standards for radiation exposure. The question arose: Is there a threshold below which exposure to radiation causes no harm? Or does the damage begin to accumulate from the very first particle that passes through us? The economic implications of the answer were enormous; they would determine the types of shields and safeguards to be engineered into the design of a power plant. The viability of nuclear power was at stake.

Inevitably, the science became contentious, with fierce accusations on both sides. The nuclear power industry is gratified to promote the idea that small amounts of radiation actually have a salubrious effect. Despite this, it happens to be true.* For adults, it may be that small doses of radiation offer a benefit of slowing the aging process that outweighs the added risk of cancer. In infants, this is certainly not true: their cells are dividing rapidly, and their DNA is more vulnerable to radiation damage. Aging is not an issue.

The epidemiology of radiation exposure is intriguing and paradoxical: cancer rates show no threshold—that is, they begin increasing with the smallest exposure to radiation. Nevertheless, other longevity factors work in the opposite direction, so that, at low levels, net life span is increased by radiation exposure.

Many animal experiments have been done demonstrating the phe-

* I do not mean to be recommending nuclear power. Good reasons to oppose nuclear power include thirty thousand years of storing radioactive waste and the potential for disasters, such as Chernobyl, Fukushima, and Three Mile Island. And without huge government subsidies, nuclear power would be uneconomic.

nomenon of hormesis for a broad variety of environmental challenges. Chloroform is a neurotoxin (used in the nineteenth century as a surgical anesthetic in low doses until too many people died on the operating table). In the 1970s, tiny amounts of chloroform were discovered to appear in toothpaste as a contaminant, and regulatory agencies demanded that the toothpaste companies study the long-term effects on health with animal experiments. The surprising result was that dogs, mice, and rats could be fed small quantities of chloroform for their entire lives, and they actually lived longer on average than control animals that were not poisoned.

Rats have been forced to slog through freezing cold water for hours every day, and the result is that they live longer. Mice raised in a sterile environment, free of bacteria and viruses, don't live as long as mice that are exposed to disease pathogens in a dirtier environment. And shocking worms with short-term exposure to heat that is not quite sufficient to kill them actually makes them live longer.

What all these experiments have in common is that the animal is treated in some way that we have good reason to believe ought to be harmful. Not surprisingly, the animal's strength and resistance rise in response. But what is completely unexpected: the body *overcompensates*, so that over the long haul, the animal does better and lives longer. This can't be reconciled with the picture we may have of a metabolism bravely fighting off challenges but being worn down in the end by accumulated stresses. Whatever mechanisms the body is using to resist stress and maintain itself against aging are not being deployed effectively in the absence of stress. This evidence forces us to see aging as "voluntary" in the sense that the body knows how to live longer, but normally it does not do so.

Living things adapt to their environments, not only over generations but (by different means) within a single lifetime. Our immune systems learn to battle the bugs to which they are exposed. When we live in a cold environment, we become cold-tolerant, and when we exercise our abdominal muscles every day, the abs (but not other muscles) become stronger. Nature demonstrates a nonverbal physiological intelligence, a silent gift for repairing damage and strengthening the faculties we use most. So we should perhaps not be surprised that when the body is exposed to a toxin or to radiation or hunger, it becomes stronger in ways that compensate and lessen the adverse impact of that particular challenge. But hormesis does more than this. The reason that hormesis is so surprising is that the body overcompensates. Living longer on less is indicative not just of repair and a gift for compensating but of a latent ability to live longer that is normally suppressed. In other words, it is clearly indicative of programmed aging.

In hormesis, in response to each challenge, the body not only mitigates the damage, it actually does better with the challenge than without the challenge. You would never predict this from any theory about resource allocation. The fact of overcompensation implies that the body has something it is holding in reserve and is not doing its best to maximize success until it is challenged.

Hormesis is an important clue about the evolutionary meaning of aging. It tells us that when the environment is rough, and many individuals are succumbing to starvation or disease or cold or heat or poisons, then aging relaxes its grip so that fewer animals die of old age. This suggests that aging serves to level the death rate in good times and in bad. Aging looks like a demographic control that keeps the population from expanding too rapidly when "the livin' is easy"; this may work because in hard times, aging can be dialed down to keep the population from declining too rapidly in the face of famine or epidemics or the like. This theory is my primary contribution to the field, and we will return in later chapters to explore it, and what it implies for arresting aging.

Instant Replay

There are three theories about evolution of aging that have become standard, and they are cited as if they are the only three possibilities. This may be true in the sense that they are the only possibilities consistent with principles of neo-Darwinism, the accepted version of evolutionary theory in the twentieth century. But there is powerful evidence against each of the three theories.

According to one, MA, aging comes from random mutations that have appeared recently, and natural selection hasn't had time to get rid of them yet. But now we know that the genes that control aging have been around for hundreds of millions of years. Natural selection would never have put up with them for so long—field studies show that aging takes a big bite out of individual fitness.

According to another, AP, aging comes from genes that offer extra strength and fertility in the early years, but these same genes cause damage and increase risk of death later on. But now we have discovered many of the genes that control aging, and some seem to have nothing to do with fertility. Indeed, some are positively ghoulish, offering no compensating benefit to the individual whatever—they are just aging genes, pure and simple.

A third theory, DS, posits aging as the inevitable outcome of a shortfall in energy needed for maintenance and repair. But this chapter has highlighted how animals deprived of food energy live much longer than those that have plenty to eat. If DS were true, we might be able to stay young forever just by gorging ourselves on food and conserving the energy it takes to get up from the couch. This is not the case.

Animals exposed to hardship often live longer than animals that have it easy. The name for this strange phenomenon is *hormesis*. It is incompatible with the idea that the body is programmed to live as long as possible. But hormesis also provides a clue about the adaptive value of aging. It seems that aging is leveling out the death rate, killing more when disease and starvation are killing less.

The stage is set for a new theory of aging, but be forewarned: any new theory may not fit so comfortably with the dominant neo-Darwinian framework on which theorists have based their thinking for more than eighty years.

FIVE

When Aging Was Young: Replicative Senescence

I strongly believe that the only way to encourage innovation is to give it to the young. The young have a great advantage in that they are ignorant. Because I think ignorance in science is very important. If you're like me and you know too much you can't try new things. I always work in fields of which I'm totally ignorant.

—Sydney Brenner

Well they're gonna tell you that everything is just dirt . . .
And that life is just to die.

—Lou Reed, "Sweet Jane"

Aging Evolved Long Ago

Life has always been vulnerable to insult, accident, and death—but not death from old age. In early times, aging didn't exist. Aging is older than animal life (half a billion years), so as long as there have been animals, they have been subject to aging. But long before that, two forms of aging appeared in microbes. Amoebas and paramecia are examples of protists, much bigger than bacteria but still only a single cell. The two most ancient types of aging that affect us—*apoptosis* and *replicative senescence*—evolved in protists.

It is curious that aging appears in unicellular life-forms. In multicellular life, it is common to hear biology teachers say, "The soma is distinct from the germ line." What that means is that there's a body (soma) made of functional cells like eyes and ears and skin and bones and muscle, separate from the reproductive cells, the eggs and sperm. The eggs and sperm have hopes of a long-term future, because they can pass their genes to the next generation. Come what may, the soma is a dead end, sacrificing itself for the reproductive cells. A husk, it is cast off as the immortal "germ cells" escape like absconding royalty to lay the foundations for a new generation. The soma is a workhorse, a slave of the germ cells. It dutifully performs its service because it is governed by the same genes as the germ line. Every cell in the body contains copies of the same genes, so there is no conflict of interest. The body's cells pass on their own genome through an exact copy in the gonads.

The separation of soma and germ line cells sets up the condition in which aging makes sense: the germ line has to be immortal, passed from parent to child to grandchild; but the soma doesn't have to be. The soma can do its job and then get out of the way. But if the germ cells ever died, it would be the end of the line, and such a species would not be around today.

According to neo-Darwinian theory, there should be no aging at all in one-celled organisms. In theory, aging begins only *after* the onset of reproduction. But in protists, reproduction is simply dividing in two. There are two identical offspring cells, clones, and no parent remains. What would it even mean for an organism to undergo aging when its entire life plan is organized around a single act of reproduction, after which the "old" cell has ceased to exist?

In fact, in his paper heralding the modern evolutionary theory of aging, the first prediction that George Williams derived was that "there should, therefore, be no senescence of protozoan clones." No doubt he thought this was a safe bet when he wrote it.

A Brief History of Life on Earth: Societies Become Organisms

The origin of life on Earth remains one of the great unsolved problems of science. It's not quite right to say that "science doesn't have a clue," because many scenarios have been outlined; but all of them contain big holes or fantastic implausibilities. One thing that is common to all the proposed scenarios: life began simple and has grown complex through widening networks of cooperation. The unit that evolves by competing with other units is what we call an "individual," but the notion of an "individual" becomes larger and broader over time.

The origin mystery invites out-of-the-box thinking, and deep thinkers from various fields of science (and science fiction) have had their say. Maybe life has always existed, built into the physics of matter. This view is less preposterous than it may seem, for several reasons. First, the oldest fossil traces of life on Earth are 90 percent as old as the earth itself. In other words, as soon the molten earth cooled enough that life was possible, there it was! Second, laboratory attempts to put together simple, self-replicating systems—systems that might plausibly have appeared "by chance" on the early Earth—have come up short. There was initial enthusiasm about this project in the 1950s, when it was reported that amino acids could be made in the lab simply by simulating lightning in an atmosphere of methane, water, and ammonia. But it takes many amino acids strung together in exact sequence to make one functioning protein, and it takes many such proteins to make a self-replicating system, and this system is far too complex and specific to have plausibly occurred by chance, ever, anywhere. Third, there are hints in the physical science of quantum mechanics that "observers" are an essential ingredient in formulating the bizarre rules that govern quantum realms. This has led some scientists to speculate that nascent life, or even conscious life, is built into the fabric of physics.

The view that life arrived on Earth originally from outer space was first seriously described by Fred Hoyle, a very smart if heterodox astronomer of the twentieth century, most famous for coining the term "Big Bang" (in an attempt to discredit a theory to which he could never reconcile). Francis Crick, who needs no introduction, was the most famous proponent for the extraterrestrial origin. Comets and meteors are continually showering Earth with matter from space. NASA estimates that one hundred tons of matter from space fall to Earth every day. "Extremophile" bacteria have been identified that can survive boiling and freezing and radiation and can plausibly hibernate as spores for the tens of millions of years that may be required to make a journey through space from one planet to another. The theory of extraterrestrial origin is, in fact, so plausible that the main argument against it is that it does not solve but only pushes back the question of how life got its start.

If life did arise from inanimate matter here on Earth, there are two main sorts of theories for how. One starts with growth and a bootstrapping metabolism and later adds individuation and division so that different pieces could originate a Darwinian process of competition among themselves. The other starts with reproduction and Darwinian competition and adds a metabolism that gradually increases in complexity over time, becoming ever more robust and efficient.

Metabolism Precedes Reproduction

The metabolic view starts with known, simple chemical systems that can "autocatalyze," or bootstrap. In other words, the product of a chemical reaction can help the same reaction along, drawing on local energy reserves and forming a fledgling "metabolism." Storms and crystal growth are familiar examples of such behavior.

Far more complex than a vortex or a crystal, life is a kind of autocatalytic chemical reaction that surfs on available free energy. Scientists look to the most universal chemical reactions and structures to find the common denominators of life. Using this method, some have suggested that life may have begun as iron-sulfur chemicals metabolized on the sides of minerals, perhaps fed by chemical energy coming from beneath Earth's crust in the early, shallower oceans.

Reproduction Precedes Metabolism

Another school of thought begins with bare molecules that could copy themselves. A barrier to conceiving simple reproducing systems of molecules was that, in all of modern life, it is DNA that stores the information, but it is protein molecules that do the work of copying DNA. It would seem that reproduction requires both proteins and DNA, and the combination is just too complex to have plausibly arisen by chance.

Then, in 1982, Thomas Cech discovered the "ribozyme." As a hybrid between enzyme (protein) and ribonucleic acid (RNA), a ribozyme is conceivably a class of molecule from which we might find an example of something that both holds information and does the work of reproducing itself. Maybe some magic RNA sequence is a candidate for the first living molecule.

The idea of an "RNA world" generated new ideas that catalyzed research in the origin of life. In labs around the world, biochemists have looked for simple RNA molecules that could conceivably pull individual nucleic acids from a dilute soup and link them together in a copy of the original molecule.

But no one has yet been able to make this work in practice. You might think that the Earth's oceans are so much bigger than a laboratory flask and that in the course of a hundred million years, something might happen that is never seen in a lab experiment that lasts, at most, a few years. But the scientist's intentional design is a vastly more efficient process than the complete randomness of a prebiotic Earth, and the fact that all our ingenuity has been unable to approach a simple self-replicating system casts doubt on the idea that this might have happened by chance. A simple ribozyme has one hundred bases stitched together in just the right order. Each base is itself a complicated sculpture of dozens of atoms, and getting one hundred of them just right is more improbable than the Earth is large. To make matters worse, the process of stitching RNA units into a chain is a reaction that goes forward only when water is removed from the system; RNA cannot form at all in open water.

It's the Cell Wall That Makes the Individual

Freeman Dyson has proposed in his "dual origin hypothesis" that metabolism and molecular replication evolved separately but then merged to their mutual benefit. Metabolism could have come before replication, which then stabilized and ensured the manufacture of the structures best able to tap ambient energy. DNA became the lighter that relit the wick of the candle of metabolism. The invention of a cell wall helped insure that the chemicals stayed put and didn't drift off into the sea, and it set the stage for homeostatic regulation so that the right proportions could be stably assured. Also with the cell wall came individuation and the possibility of competition, which was to become the driving motive for new strategies and ever-growing complexity.

Over evolutionary time, species don't just branch off, but organisms also come together to form individuals at higher levels. Self-reproducing molecules joined forces with other molecules that were able to reproduce *each other* with mutually higher efficiency. In a few rare but profoundly transformative events, a parasite cell would merge with a host to form a symbiotic combination. Species that were once mortal enemies—predator and prey, or parasite and host—first learn to coexist and then to help each other, and then they may grow so close that they can no longer live separately. Lichens are combinations of algae with fungi that have reached this stage.

And there is yet a stage beyond where the former enemies merge their genomes into a single species, a single cell, and the separate origin is not at all obvious without close inspection. We saw in chapter 2 that mitochondria evolved from invading bacteria that learned first to refrain from killing their host cells and then to share their energetic resources for all the cell's metabolic purposes and finally to subjugate their own metabolism under direction of the nuclear DNA. Similarly, chloroplasts are tiny islets of photosynthesis within plant cells, but once upon a time, they were independent cyanobacteria cells that were ingested for food by a larger eukaryotic cell.

The eukaryotic cell is no mere aggregation of mitochondria and chloroplasts but a diverse system of functioning parts, many of which were independent organisms at one time. Two-thirds of the way through life's journey on this planet, the first eukaryotic cells appeared, miracles of complexity, typically a million times larger than bacteria or archaea. How these diverse colonies came to be integrated into a functioning unit is a subject of human speculation, but alas, we came along too late to witness the event. After several hundred million additional years, the clusters of eukaryotic cells first differentiated into specialized tissues, working together as a multicelled organism. Nature's most spectacular example of cooperation is, of course, the one that is so familiar to us that we never give it a moment's thought. The familiar macroscopic biosphere of plants and animals is but the peak of a pyramid, whose base is the yet-uncounted diversity of bacteria. The biomass of bacteria in soil, in oceans, and deep in the earth exceeds all the plants in the world, which, in turn, have far larger biomass than the insects, which are by far the dominant animal form. All the fish and birds and mammals that we conjure when we imagine "animal life" constitute less than 1 percent.

Very soon after larger life forms appeared, communities of animals learned to cooperate. Of course, wolves hunt in packs, but more surprisingly, moray eels and grouper fish communicate with sign language to hunt for prey that they could not catch separately. Groupers can win a chase in open water, but when smaller fish hide in coral cavities and crevices, the eel can go in after them, trapping the poor fish between two hungry mouths. Coevolution between species is too common even to offer a fair sample, but wasps and fig trees are one spectacular example, with several hundred species of wasp, each exquisitely adapted to pollinate a particular species of fig and to feed on its nectar. Whole ecosystems of crustaceans are adapted just for recycling a whale's carcass half a mile below the ocean, and many of the bacteria in your gut could not live anywhere else.

"Eusociality" is a word biologists reserve mostly for social insects, cooperating so tightly that only one queen in a hive gets to pass on her genes, and that's perfectly okay with the thousands of worker and drone bees inside. It used to be said that this level of cooperation was only possible because the queen shared 50 percent of her genes with the inmates of her colony; but in recent decades, cases have been discovered in which the slaves have been recruited to work for a queen without genetic relationship.

The anthill or beehive can be viewed as a "superorganism," with many of the properties of an individual. Certainly, in terms of natural selection, there is no individual competition of one ant against another but competition instead at the level of colony against colony. The concept of a superorganism was parodied with Woody Allen's zany deadpan in the movie *Antz*, twenty years after Douglas Hofstadter punned on the name of "Aunt Hillary" in his fictional/scientific masterpiece, *Gödel, Escher, Bach*. Termites are a "fractal" example of a superorganism, because not only do they live in eusocial colonies but each individual termite is a superorganism in its own right, dependent on a community of gut bacteria.

Ecosystems are the highest level of organization, and (though this is controversial) I have no trouble acknowledging evolutionary competition on the level of competing ecosystems. Successful ecosystems grow and expand, encroaching on the territory where less robust ecosystems have become dysregulated and vulnerable. Ecosystems are almost certainly as old as life itself, as all life is interdependent, and no single species can survive without a web of other species for support. This is obvious in the case of animals but hardly less obvious in the case of plants, which cannot live without the CO_2 exhaled by animals or the nitrates delivered to their roots by nitrogen-fixing bacteria. We live on a symbiotic planet.

Is it sound science to describe "Gaia" as an entire planet integrated and organized with a global physiology? Many great minds have found inspiration in this idea, including James Lovelock, Lynn Margulis, and George Wald.

Darwin described the target of natural selection not as strength or speed of reproduction or dominance in a battle of wits or brawn; rather, he used the word "fit." An individual or a colony or a species is successful in nature's contest to the extent that it fits in with others in its ecosystem to make a robust, thriving community.

In this context, the standard neo-Darwinian dogma that says "competition is always individual versus individual" looks like an oddity, an arbitrary article of faith rather than a law of nature. The definition of "individual" has changed several times over evolutionary history. Today's group becomes tomorrow's individual. It seems plausible that competition and cooperation are happening always and concurrently at several levels. This is the principle of "multilevel selection," a paradigm (introduced by my mentor, David Sloan Wilson) that is making its bid to replace the selfish gene.

Sex

The exchange of genes is at least as old as cellular life. I'm sure you'll agree that sex was one of life's all-time best inventions. Sex ties communities together and assures that evolution is not hijacked by selfish individuals. There is always the danger that community members can turn selfish, behaving like parasites and stepping on the toes of others to get ahead. They might do well within the protected environment of the community, but if they were allowed to *become* the community, they would no longer have other toes to step on. A community of selfish individuals is not much of a community, and it finds itself at a disadvantage competing with more cooperative communities. Sex tied to reproduction ensures that organisms must seek one another out as mates, and this predisposes organisms for sociality. Sex as it exists in animals, where it is mandatory for reproduction, helps thwart the selfish gene. When there is a single gene pool shared by the community, genes are compelled to work well with others, or they don't get very far.

Although sex is important for the community, individuals—especially successful individuals—are always, in the neo-Darwinian perspective, tempted not to share their genes. Those that are reproducing fastest are taking over the population with their progeny. Why should they share the keys to their success? If evolution were entirely a matter of selfish genes, then the selfish genes would keep success within the family and not share the secret of their success with other lineages. The reason that sex became a prerequisite for reproduction is to compel successful individuals to share their genes.

Like wildcat capitalism, selfishness works great for a while but ultimately leads to disaster. "Dog-eat-dog" is not a long-term recipe for evolutionary success, while successful community building is a proven strategy that has worked at ever-higher levels of organization through life's history.

Without sex, evolution might not work nearly so well. Sex is important for the long-term viability of life. And yet, in the short term, from a selfish gene perspective, there is a constant temptation to avoid sex and just reproduce by cloning. Sex is theoretically ever in danger of being lost, and it is vitally important to the long-term health of most biological communities that sex be maintained and that every individual participate.

The all-time most successful animal species are all social insects. There are more social insects than anything else. And even though each one is so tiny, the total biomass of all the ants, termites, bees, and wasps in the world exceeds the total biomass of all mammals. Put another way, there are about seven billion humans on the planet, and *for each human*, there are about seven billion ants. Humans have only recently arrived at their position of planetary dominance, and our exploitation of other life-forms is dangerously unsustainable. Ants have been around the block a few times and by now are thoroughly and stably integrated into the world's ecosystems, from Amazon rain forests to our kitchen cabinets.

Ants do not dominate the earth's biomass, however. That distinction belongs to bacteria.

Avoiding the Sin of Chastity

Sex and No Sex are two separate evolutionary games with separate sets of rules. The No-Sex game, comparatively, is hard-edged competition, a clone fest, where winners take all. In the Sex game, by contrast, almost everyone goes home with a prize, though some prizes are bigger than others. There is much more experimentation in the Sex game than with No Sex, as genetic recombination is installed at the species level so there's much more innovation in the long run. Evolution with Sex can go places that No Sex may never find. There's little question but that the Sex game is a better game, a more interesting game. For the long haul, there's good reason for throwing in your lot with the Sex community.

But at any given moment, an individual who is a strong competitor may be tempted, if the physiological opportunity is there, to play the No-Sex game, all or nothing. If you're playing all out and I'm playing soft competition, then it's likely you're going to roll right over me. If one individual only is playing the No-Sex game (cloning itself) while everyone else is playing the Sex game, the individual without sex is poised to wipe out the competition in the short run—then stagnate in the long run.

The game of Sex is a superior game, but only if everyone plays by the rules. Where the Sex game is played, it is vulnerable to invasion by No Sex—that is, by the evolution (or perhaps devolution) of cloning, which produces more selfish genes but at the cost of diversity and sustainability. The Sex game needs to protect itself from encroachment by the No-Sex game, or Sex will die out, experimentation will languish, and exploration of new niches will slow to a crawl. How can a sexual species protect itself from that mutant individual who abjures sex and challenges the community, winner take all? A billion years before Augustine and Gandhi engaged in lifelong struggles with the temptation of sex, evolution struggled with the opposite temptation—the temptation to forgo sex.

The Carrot and the Stick

So evolution has been faced with the dilemma: Evolving cooperative communities is potentially a winning strategy, and sharing of genes is a deft way to tie together the fate of a community and make selfish behaviors less profitable. But how is sharing of genes to be enforced? The danger is that individual selection, acting quickly and efficiently in the short term, can boost the prospects of anyone who has a temporary, contingent advantage in individual fitness and who refuses to share that advantage with others.

Natural selection has come up with two (rather heavy-handed) means for enforcing gene sharing, which I call the Carrot and the Stick.

In sexually reproducing animals (that's you, me, and the cockroach), evolution uses the Carrot. Sex feels good, and there is a powerful instinctive urge reinforced through our nervous systems. But that's just the superficial Carrot. The deep Carrot is the opportunity to reproduce. In most animals, sex has been tied so tightly to reproduction that the two processes are absolutely intertwined. Not only is it impossible for the individual to reproduce without sex, but it is extremely difficult to evolve backward to a capacity for clonal reproduction. It has happened, for example, in aphids (greenflies), whiptail lizards, and (in plants) dandelions, but it is rare and difficult.

We think of sex as part of reproduction, and that's just the way nature wants it. We've been brainwashed to think that sex and reproduction go together like a horse and carriage, so to speak. But in fact, sex is about sharing genes, and reproduction is the process by which an organism creates copies of itself, and there is no necessary connection at all. Once upon an earlier time in evolution, sex and reproduction were entirely separate, and in many protists, the functions remain largely separate to this day.

Aging in Protists: A Death Sentence for Refusal to Share Genes

A few pages back, we began talking about aging in protists, and then we detoured for a whirlwind history of life. Now we are ready to understand what aging in protists means and how it evolved.

It was in protists that natural selection chose the Stick. Your DNA or your life! Share your genes, or you die. Ciliates are a protist phylum that goes back at least 580 million years (and probably much longer) and includes paramecia. It is probable that animal life evolved from ciliates.

Sex and reproduction are entirely separate functions. In ciliates, reproduction is by cloning (mitosis), or simple cell division; and gene-sharing is by a process called conjugation. The two processes are metabolically independent, so there would be nothing to stop a successful individual paramecium from eating and cloning, eating and cloning, until its descendants took over the colony in a monoculture. So natural selection, wagging her finger, has erected a barrier against this. Each time the paramecium reproduces, it loses a little bit of DNA from the tail of its chromosomes, called a telomere. With each division the telomere shortens, until the chromosome becomes unstable and can't function. The cell languishes and dies. The cure for this malady is an enzyme called telomerase. It acts as an antidote to death, restoring the telomere. The punch line is that telomerase is kept locked up in a DNA vault. Evolution has prevented the paramecium from getting to the cookie jar except when it has sex. The restorative enzyme is only available during conjugation.

The result is that any clonal line can go on reproducing for a few hundred generations, but then it runs out of telomere and the whole lineage will die out unless they share their genes via conjugation.

This is the earliest known form of aging, from a time when all life on Earth was microscopic. In retrospect, we may say that aging evolved for the purpose of protecting the community, enforcing the imperative that selfish individuals must share their genes. The biochemistry of aging in single-celled protists carried forward and is substantially identical to cellular aging in your cells and mine. This continuity is a clear indication of the evolutionary meaning of aging.

The word "cilia" comes from the Latin for "eyelashes." Ciliates swim by waving thousands of tiny hairs (cilia), with which their tiny bodies are covered. Ciliates show the same complex internal structure of microtubules (nine pairs surrounding a central pair) that can be found in the follicles of the fallopian tubes of women and the sperm tails of men. The fine structure shows they are ancient and have been conserved over a vast swath of evolutionary time. Lynn Margulis, a highly creative biologist who lived to see some of her wilder ideas validated, argued that the cilia themselves came from another ancient symbiosis, perhaps spirochetes.

How Replicative Senescence Was Discovered

The loss of a little bit of telomere with each replication is called *replicative senescence*, and it is an early form of aging that applies to individual cells. It is only in the last fifty years that biologists have understood that individual cells can age. Before that, aging was thought to be a property of systems of cells, not of the cells themselves. One man was responsible for convincing the science community otherwise, and his name is Leonard Hayflick.

As a microbiology grad student in the early 1950s, Hayflick learned about the longest-running experiment in biological history. Alexis Carrel had kept a cell culture alive, growing and replicating for thirty-four years, from 1912 to 1946. This proved that cells themselves were immortal and that aging must take place at a higher level within the organism as a whole.

Carrel was a French doctor who pioneered organ transplant operations. He moved from Paris to Chicago, where the Rockefeller Institute recruited him. In 1912, Carrel won the Nobel Prize in medicine for figuring out how to join blood vessels so that they would grow together and feed the transplanted organ. On a return visit to Paris, Carrel gave a presentation about his new interest: growing animal cells—including human cells—outside the body in lab conditions. He crowed to his compatriots about the superior facilities and research environment in America. Incensed, the Europeans challenged him to demonstrate his heretical ideas.

Carrel took the dare and, upon his return, set about demonstrating rigorously that cell cultures could be grown in a test tube. He sent an assistant to learn the technology for keeping them alive and started his own culture with embryonic cells from a chick heart. The experiment proved successful, and he wrote about it for two years, adding to his fame and influence. Carrel, as it turned out, was articulate and charismatic; this combined with his scientific success turned him into a minor celebrity and an unchallengeable authority.

Carrel passed off his chick cell culture to a colleague, Arthur Ebeling, who maintained the experiment for thirty more years. Though he had let the experiment out of his personal control, Carrel remained interested in the results and soon became an advocate for the view that individual cells could divide and grow indefinitely. Over the decades, it became a piece of the scientific canon: animals may get old, but the cells of which they are made have the capacity to grow and divide indefinitely.

Scientific canons are difficult to oppose. Other researchers were growing cell lines in the 1940s and 1950s and saw different results.

They published their findings but always attributed the death of their cell lines to special conditions and did not challenge the orthodoxy. A theory (a myth, actually) grew up that kept scientists from having to confront the contradiction: that something in the chemistry of the culture became poisoned with age, and this would prevent the culture from continuing to grow, even though the cells themselves were immortal.

And so we return to Leonard Hayflick. Just a few years out of grad school, he was recruited to head the cell culture laboratory at the University of Pennsylvania's Wistar Institute. There he designed an experiment to compare cancer cells to normal cells, and he didn't think to question the scientific results he had inherited: cell lines could be grown in vitro without limit.

But he soon encountered difficulty with his cell lines. The normal cells (not the cancer cells!) would slow down and die after a year or so, corresponding to about forty cycles of growth and division. At first, he assumed that his lab technique was to blame. But after a few years of refined experiments, he was convinced that cell aging was not an artifact. At this point, he switched gears in his research. Putting cancer aside, he asked the basic question, how could he demonstrate cellular aging unambiguously?

His idea was to grow old cells and young cells in the same culture. If Carrel's claim was correct, then the poison from the old cells ought to kill both. But if the problem was endemic in the cells, the young cells might continue to grow even as the old ones died.

But how could he tell the two cell lines apart? Remember that this was 1959, before modern techniques of biochemistry or DNA sequencing. Hayflick solved the problem with cells derived from men and women. Even though the biochemical details could not be discriminated, the difference between a full-size X and a stubby Y chromosome could be easily discerned under a microscope. Hayflick grew cells from a man's skin until they got old and their reproduction slowed. He added fresh skin cells from a young woman.

In a short time, there were only female cells in the culture, continuing to grow and to replicate. The result was clear: the "poison" theory was dead. Somehow the cells knew how many times they had divided, and they slowed down with age. Once this dam had broken, it was not long before multiple labs replicated the result. A scientific doctrine established for decades fell quickly. Hayflick repeated his experiment using cells aged in vivo; he began his colony with skin cells drawn from older humans and younger humans. Cells from the older people had a shorter lifetime in the laboratory before they showed signs of aging. This confirmed that the phenomenon he had discovered was not some laboratory oddity: within the body, cells are aging measurably. This reopened the possibility that the aging of an animal has its roots in processes at the level of the cell.

How Did This Happen?

For anyone seeking reliable truth from the scientific establishment, this is a cautionary tale about charisma and the power of groupthink. The need for replication as an essential step in the scientific process had been honored, yet so powerful was the scientific consensus that contradictions in the new results were explained away, sidestepping the need to challenge an established dogma. Scientists are human.

Still the question remains: How could Carrel and Ebeling—good and careful scientists both—have obtained the results that they did over such a long period of time? Hayflick was polite enough to suggest in his original paper that Ebeling's medium had inadvertently become contaminated. The nutrients that were added daily to the cell culture were derived from fertilized eggs. If occasionally the nutrient broth was not properly sterilized before being mixed into the cell culture, that would provide a fresh supply of embryonic cells that could explain the apparent longevity of the cell culture.

We'll never know for certain, since "the ultimate effects of the aging process have made it impossible for Carrel to respond in his own defense." But there is a relevant anecdote from a contemporary scientist who visited Carrel's labs in 1930. Ralph Buchsbaum claimed a lab technician told him that periodically the cell line would be discovered dead, and the technicians, thinking that they had made some mistake, added fresh cultures to cover their derrières.

Telomeres and Cellular Senescence

Although Hayflick discovered that cells could not go on dividing forever, he did not yet know the reason. What happens to cells as they reproduce that eventually slows them down? How does a cell keep count of how many times it has reproduced?

In the mid-1970s, another young postdoctoral researcher, Elizabeth Blackburn of Yale (now at UC–San Francisco), found direct and compelling answers to both these questions. She was working not with human or animal cells but with single-celled paramecia. Chromosomes are long strands of DNA found in every cell nucleus, with hundreds of millions of nucleic acid subunits, the bases T, A, C, and G. The order of the subunits contains blueprints for manufacturing the giant protein molecules that the cell will need to live and do its job. When the cell divides, the chromosome uncoils its double helix, and each strand creates a new mate. There is a molecular engine, DNA replicase, that crawls down the length of the chromosome and collects new bases T, A, C, and G to create a matching half of the chromosome, and this is how a single set of chromosomes becomes a double set, for two daughter cells.

Here is what Blackburn discovered: the replicase enzyme passing over the chromosome has trouble at the end of its mission. It runs out of room to secure itself and cannot properly copy the last few hundred base units. Micromechanically, the reason is that the molecule that copies the DNA actually attaches to the chromosome and slides along as it copies. When it gets to the end, it can't copy the part of the chromosome that it is sitting on. The result is that every time a chromosome is copied, it becomes a bit shorter.

Well, how can that be? Clearly the cell can't afford to be continually losing information. The solution, also discovered by Blackburn, has two parts. First, there is a buffer of DNA at the end of every chromosome that carries no meaningful information. This is called a telomere, and it is just a repeated pattern of bases—TTAGGG, repeated over and over again, thousands of copies strung end to end. In a functioning chromosome, the telomere tail automatically folds over on itself to give the DNA a neat termination that isn't chemically reactive so that the double helix can't come unwound. These days, Blackburn likes to introduce her public lectures with slides showing the aglet at the end of a shoelace, preventing the lace from raveling. (I've been tying my shoelaces for more than sixty years, but it wasn't until I heard Blackburn's lecture that I knew the word "aglet.")

The telomere protects the information-carrying part of the DNA and provides a disposable region that can be transcribed or not, without loss of information. The telomere is typically tens of thousands of bases long, and it continues to function as an end cap even if a few hundred or a few thousand bases are lost in multiple generations of copying. (The chromosome in its entirety is far longer, with tens of millions of base pairs.)

The second part of the solution: How are telomeres restored? Having a buffer of meaningless DNA is a temporary expedient, but sooner or later, the buffer must be rebuilt. This, too, was anticipated and solved by Blackburn in her early work on the biochemistry of the telomere. There is an enzyme that she named, logically enough, telomerase, the function of which is to add copies of TTAGGG to the end of a chromosome. Normally, DNA is copied to messenger RNA as the chromosome sends messages out to its protein factories (called ribosomes), giving instructions for what proteins to make and how to make them. This order of nature was immortalized by Francis Crick, who named it the "central dogma" of molecular biology. But like other dogmas, it eventually encountered exceptions and was forced to be less dogmatic. Rarely, RNA is transcribed backward into DNA. Rotavirus is an example: it is made of RNA, but it is translated with reverse transcriptase inside human blood cells and becomes part of the DNA.

Telomerase uses this same trick: it contains a short piece (the six bases) of RNA as part of its molecule and uses this RNA template over and over to extend the telomere tail (even though the telomere is made of DNA). Telomerase is the answer to the question, how is it that the telomere grows back the length that it loses in cell replications?

If the telomere were to shrink to zero, the chromosome would start to lose real (coding) information with each additional replication. But it never gets that far. When telomeres are starting to get short, the chromosomes become misformed and unstable. When there are still thousands of bases left in the telomere, the cell already senses that it has a short telomere, and it goes into a senescent mode. At first, metabolism is slowed, and as the telomere shortens further, the cell becomes inactive and ceases to divide further. It may begin to spew poisons that damage younger cells around it. It may commit suicide via apoptosis, a process described in the next chapter; but in any case, it has ceased to have an active biological function, and in one way or another it will die.

Leonard Hayflick

Eighty-six years old at the time of this writing, Len Hayflick remains a revered and active member of the aging research community. He is fiercely independent, kindhearted, a bit irascible, and a self-styled curmudgeon. Though he discovered one of the oldest mechanisms of programmed death in cells, he does not himself believe that aging is programmed in humans. In fact, he has written in support of the "wearing out" theory that we have decried in chapter 1 as a misreading of thermodynamic theory.

Surrounded by enthusiastic colleagues, Len is skeptical about the prospects for antiaging medicine and dubious about the effect on society of prolonging life. In this, he joins a venerable tradition of those who have warned against interfering with long-evolved nature.

How Cell Senescence Contributes to Aging in You and Me

In paramecia, telomerase isn't available during normal cell replication but is only expressed during the comparatively rare event of conjugation. In people, telomerase is mostly unavailable during our lifetimes but only comes out in the embryo. One of the first things that happens after sperm joins egg and an embryo begins to form is that the DNA is reprogrammed for a fresh start, and part of that process deploys plenty of telomerase so that the embryo begins life with long telomeres. Typically, a human embryo might begin its development with twenty thousand units of telomere, but the stem cells are dividing so rapidly in the womb that by the time a baby is born, only ten thousand units remain. These ten thousand must last us the rest of our lives, and the telomeres in our cells are continually shortening over a lifetime of cells dividing for growth and repair.

After the discovery of cellular senescence and its explanation via telomere dynamics, it became clear that cells in our bodies must also have a limited lease on life. But everyone assumed that this could have nothing to do with aging. According to the prevailing paradigm, the body is doing all it can to resist aging. That means that we could never die from something as simple as short telomeres. If that ever began to happen, then evolution would just provide us with longer telomeres or with a lifelong supply of telomerase, as mice have.

But we *do* die simply for lack of telomerase. The proof was demonstrated by Richard Cawthon in 2003. People with short telomeres have a much higher risk of dying (and especially of heart disease) than people the same age who have longer telomeres. If we just had more telomerase, we would have longer telomeres, and we would live longer. Since 2003, the association between telomere length and aging has been confirmed in many birds and mammals.

Stem cells with short telomeres accumulate as we get older, and they hurt the body in three ways. First, they stop reproducing. It's the stem cells' job to regenerate the tissues that are damaged by normal use. The stomach lining, constantly exposed to hydrochloric acid, is being replaced constantly. Blood cells and skin cells turn over every few days. But as the stem cells slow down due to telomere senescence, there is less replacement. Second, the cells with short telomeres have chromosomes that tend to come unraveled and become damaged. This can lead the active stem cells to a cancerous condition. And third, senescent cells emit signals that cause increased inflammation all through the body. Inflammation is one of the primary ways that the body destroys itself with age (more on this in chapter 9), and senescent cells are pouring gas on the fire.

Richard Cawthon and the Role of Telomeres in Aging

"It can't be so simple" was the conventional wisdom. If our bodies were simply getting old and dying for lack of telomerase, our bodies would produce more telomerase and solve the problem. It was known that telomeres get shorter over the human lifetime but that most cells have enough telomere length to last us through. That was all we needed. Our bodies are smart enough to make telomeres just long enough to keep our cells healthy through a normal lifetime, but not longer. Keeping the telomeres trim was thought to be the body's way to thwart the runaway replication of cancer cells.

It was obvious that our bodies could not be dying from telomeres that are too short, so obvious that no one thought to test the idea. Besides, testing human populations for mortality takes a decade or more, and costs a fortune.

Richard Cawthon was a biochemist at the University of Utah in Salt Lake. He wasn't convinced that there is no cost to short telomeres, and he had an idea how the question could be answered quickly and without the expense of a long-running epidemiology study. First, the biochemistry: he figured out how to measure telomere length of a sample of just a few cells. He did this by replicating the DNA many times over, cutting the DNA into fragments, and then adding a reagent that sought out the pattern TTAGGG and stuck fast to it.

A second innovation was that, instead of testing people's telomere length and then waiting for some of them to die, he thought of a way to use historic human data. In Utah, many people stay put for a long time. There were records from people who had donated blood twenty years earlier, preserved in hospital refrigerators. Many of those same people had stayed in the area, living and dying in Salt Lake City.

In the study that Cawthon published in 2003, he measured telomere length from blood cells of just 143 people, all of whom had been sixty years old when the blood was drawn. He looked to see if a relationship existed between the lengths of their telomeres and the subsequent fate of each individual. He found a powerful relationship between life expectancy and telomere length. People with shorter telomeres had far higher death rates from infectious disease and from heart attacks. They did not have lower cancer rates than people with longer telomeres.

If the prevailing theory about cancer were correct, then there should be no net relationship between telomere length and life span. What he found instead was that people with the shortest telomeres were dying twice as fast as people with the longest telomeres.

In the years since, this relationship has been confirmed in several other studies of humans and also in several kinds of mammals and one bird species. A very large Danish study in 2015 was able to separate short telomeres from all the standard risk factors like weight and smoking and cholesterol and to show that short telomeres were more closely associated with risk of death than any of them.

Telomeres and Cancer

The telomerase gene is available in every cell, and yet it is kept locked up, unexpressed, epigenetically suppressed while the cell dies for lack of it. This looks inescapably like programmed aging. What do mainstream biologists have to say about this situation? They say that suppressing telomerase is a way to avoid cancer. They argue, reasonably enough, that cancer cells could not grow uncontrollably and threaten the body the way that they do if they did not have a way of escaping the Hayflick limit—the number of times, about forty, that the body's tissue cells can divide. And indeed, almost all cancer cells have epigenetic modifications that allow them to express just enough telomerase to keep going.

Biologists who believe in the AP trade-off theory of aging say that the reason that telomerase is normally under epigenetic lock and key is to protect the body against cancer. A shortened life span is the price we pay for protection against cancer when we are young.

But experiments with lab animals don't support this theory. The problem is that the immune system provides much better protection against cancer than short telomeres can. When we are young, cancer is rare because our immune systems are robust and protect us well against incipient cancer. But as we get old, the immune system is weakened by short telomere disease. The stem cells that create new white blood cells slow down because their telomeres are getting too short.

It gets worse. Chromosomes with short telomeres become unstable and are prone to becoming cancerous. And before the telomere becomes so short as to actually kill it, the cell goes into panic mode and sends alarm signals out through the body, calling for inflammation. Inflammation is the biggest reason that cancer risk goes up as we age.

The bottom line is that short telomeres cause much more cancer than they prevent, so the AP interpretation of telomerase rationing makes no sense.

The Big Picture: The Same Old Same Old

Thus the circle is closed. Cellular senescence evolved half a billion years ago to enforce gene sharing, to keep selfish individuals from dominating the gene pool. And even today, the same mechanism and the same biochemistry serve the same function. Cellular senescence contributes to whole-body aging. Aging, in turn, kills off the most successful individuals, the only ones to survive to old age, with the result that they do not dominate the gene pool, that there is turnover in the population, and that the young have a chance to grow up and flourish in a less crowded niche.

This is the cycle of life. The individual is born and dies, but the community goes on. This is the stuff of which myths are made.

Instant Replay

Neo-Darwinian theory says that aging shouldn't exist in one-celled protozoans. But the protozoans haven't read the theory book. There are two forms of aging in protozoans, and because they are so old, these are powerful hints about how aging evolved and what adaptive purpose it serves.

One is called cellular senescence, or replicative senescence. Using the telomeres, tails on the ends of each chromosome, cells keep count of how many times they have copied themselves, and after a fixed number of copies, the telomeres become short, and the cells get tired and die. Replicative senescence continues to this day as part of the biochemistry that makes us old, a means to an end. Cells of older people have shorter telomeres, and this slows down repair and also poisons the body.

Aging in protists evolved as a kind of policing mechanism, forcing the most successful individuals to share their genes with the community. This was as essential then as it is now for keeping the community diverse so that it can adapt to change and continue to evolve. Aging continues to work in this way in communities of higher organisms, keeping the community diverse, making sure that even the most successful individuals don't live long enough to dominate the community with their genes.

SIX

When Aging Was Even Younger: Apoptosis

My heart leaps up when I behold
A rainbow in the sky:
So was it when my life began;
So is it now I am a man;
So be it when I shall grow old,
Or let me die!
The Child is father of the Man;
I could wish my days to be
Bound each to each by natural piety.

—William Wordsworth, 1802

The alternative to thinking in evolutionary terms is not to think at all.

—Peter Medawar

Can a Cell Be a Good Samaritan?

The neo-Darwinists who have dominated evolutionary theory take a dim view of altruism. Theory tells them that altruism shouldn't exist and that any appearance of altruism is really an illusion. So when Valter

Longo as a young doctoral student at the University of Southern California claimed that yeast cells were sacrificing their very lives for the good of the community, journal editors rejected the research report that he submitted and sent him back to the lab to double-check his results.

Some people quietly accept rejection, resigning in the face of obstacles; others rise to a challenge and redouble their efforts, thriving on the provocation. The journal editors underestimated Longo, his acumen in the lab, his boundless energy, and his perseverance. The end result of putting him through his paces, requiring layer upon layer of evidence, was a bulletproof case for altruistic suicide in yeast cells featured prominently in the journal *Nature* in 2004.

Discovery of Cell Suicide

It was first observed in the 1840s that cells sometimes kill themselves. Ordinarily, cells will fight approaching death with every means at their disposal. If they are short of food, they will take emergency measures, beginning to digest their spare parts. If they are short of oxygen, they will move into an anaerobic mode for energy generation. If they are poisoned, they pump the poison out of their cytoplasm as fast as they can. By the time they die, they are battle-worn and show multiple signs of damage.

But apoptosis, cell suicide, is different. In apoptosis, the cell makes an orderly plan for its own demise. It chops its own DNA into pieces that can no longer direct the cell's metabolism. It burns (oxidizes using hydrogen peroxide) the highly refined proteins that once regulated its chemistry. Far from resisting death, the cell is working in an efficient and orderly manner to shut down its own metabolism and turn itself into food for its neighbors. The last thing that the cell does is dissolve its cell membrane, scattering the aqueous equivalent of its own ashes.

Apoptosis can be triggered in a process of self-policing, as when a cell detects that a virus has infected it and it falls on its sword rather than give the virus more opportunity to grow and spread. Precancerous cells figure out that they are precancerous and eliminate themselves. And when the young body is forming itself, much of its shape is sculpted not with creation but with destruction, as the unwanted in-between areas are eliminated with apoptosis.

A somatic cell might thus fall on its sword for the good of the body. But could an independent living cell give its last full measure of devotion for its neighbors?

From Murder to Suicide: The Taming of Cell Executioners

In chapter 1, we first met mitochondria—tiny organelles stippling the inside of each eukaryotic cell and generating its energy. Electrochemical energy is necessary for everything that a cell does. Eukaryotic cells are able to burn sugar with oxygen and supply that energy in a form that the rest of the cell can use. Once upon a very long time ago, there were cell nuclei, but not yet the full metabolic machinery of the modern cell, and ancestors of mitochondria were invading bacteria. A deadly infection, they entered the cell, burned its sugars, and used the energy to make copies of themselves. They were parasites, adventitious, opportunistic, greedy, and selfish.

Of course, mitochondria carried their own DNA, which means their loyalty was not to the cell's nucleus but to their own agenda. Like marauding bandit armies, they would commandeer the cell's metabolic machinery to make copies of themselves. Eventually, their wild profligacy would kill the cell, and mitochondria would spill out into the sea, looking for other cells to infect.

But in this relationship lay the seeds of cooperation. The mitochondria's livelihood depended on stealing from the cell. The more efficiently they were able to steal, the faster they could grow and reproduce. But like all parasites, the mitochondria were biting the hands that fed them. The more efficiently they were able to steal, the faster they killed their host cells and put themselves out in the cold ocean, where the great majority of them must have died in search of other cellular parties to crash.

Evolution in the short term tends to pit parasite against host, but in the long term, these two often meld into peaceful coexistence. Parasites quite generally evolve to become less virulent over time for just this reason: it is not in their selfish interest to kill their host but to keep it alive so it can be milked more effectively. It is a further step to actually support the host, to provide services that enhance the host's ability to thrive and provide more resources for the parasite to steal. Evolved symbiosis is a central theme in biology.

Bacteria evolved from virulent parasites into mitochondria. There are hundreds of mitochondria in each eukaryotic cell and thousands in cells that use energy intensively. All the chemical energy that powers plants and animals today is processed through mitochondria.

Can you imagine a wolf devouring a young deer, not digesting it completely, and becoming a new sort of animal with all the strongest traits of the wolf and the deer together and some new qualities as well that grew out of a synergy between the wolf and the deer? It sounds like a fairy tale. But at many points during evolutionary history, organisms with completely different backgrounds, different genetics, lifestyles, strengths, and weaknesses somehow merged to form a new organism. Each time, the new creature was more complex by a quantum leap, more adaptable, and able to do things that none of its progenitors was able to do. Sometimes one ate the other and didn't digest it completely; or else the junior partner started as an invading parasite that learned first to avoid killing its host and then, ever so slowly, to work in support of the host on which it depended.

It was more than two billion years ago, about half the age of the earth, when the first eukaryotic cell was formed. The Spanish microbiologist Ricardo Guerrero makes a good case that this event was more revolutionary than anything that has happened since. Before eukaryotes, there were only bacteria and (equally tiny) archaea. They had not much internal structure, and their genes were (and still are) in little loops called *plasmids* that circulate through the cell.

Eukaryotic cells are about a million times bigger, highly structured, and tightly regulated and integrated. All their genes are in chromosomes in a cell nucleus that runs the show. There are hundreds of different units within the cell, *organelles*, each specialized in doing some task that the cell needs. These tasks include generating energy, manufacturing various proteins, defending against invaders, sensing the environment, changing the shape of the cell, and locomoting, which is performed by waving little cilia like tiny oars in the water. There are networks of roads that transport important custom-manufactured molecules to the place where they are needed, and there are destination tags on the molecules that tell the machinery where to take them. The new complexity created a niche for leadership, and the cell nucleus took on that role, a kind of central government that evolved in the aftermath of the multispecies merger.

All familiar largish life-forms, animals and plants and fungi, are made of eukaryotic cells, but even the one-celled eukaryotes (protists) are much larger and structurally more complex bacteria than the bacterial assemblages from which they came. Bacteria and archaea are, however, metabolically far more diverse, and ultimately tougher, than the larger forms we are perhaps too quick to consider "higher."

Darwin believed that all evolutionary change occurs gradually, incrementally. A century later, Steven J. Gould popularized the idea of *punctuated equilibrium* to describe very quick changes in evolutionary history, in which the fossil record shows a discontinuity between what came before and what appeared afterward. Through a long career in the biological sciences, Lynn Margulis put new focus on major evolutionary transitions and put forward the theory that it was a process of mergers and acquisitions that propelled evolution forward in startling leaps. *Symbiogenesis* is the concept that she bequeathed to us.

What Has This to Do with Apoptosis?

As a vestige of their evolutionary past, mitochondria retain a snippet of their own DNA, and they have their own reproduction cycle. Mitochondria die, and new mitochondria are cloned from the old, even as the cell goes about its metabolic business. When you exercise, your muscles get the message that they need more energy, and they in turn signal their mitochondria to proliferate. Mitochondria have much shorter lifetimes than the cells where they make their homes. They burn themselves out and need to be recycled.

Mitochondria also retain from their bandit past the capacity to kill the cell. They don't do this for selfish reasons anymore. The signal that initiates their murder-suicide originates in the command center of the cell nucleus. But the destruction is carried out with some of the same simple and highly reactive chemicals that are part of the mitochondrial energy-generating cycle. Hydrogen peroxide, H_20_2—water with an extra oxygen atom—is a common antiseptic you can buy in the drugstore. It is also a mitochondrial product. In excess quantities, it becomes both signal and the means of destruction. Peroxide is too strong a poison to live in the cell's cytoplasmic soup, and so the mitochondria have efficient chemical means to mop up their peroxide with free radical scavengers as fast as it is produced. When the signal for self-destruction is received, mitochondria respond by creating peroxide in much larger quantities. The "contract killing" begins. The peroxide spreads through the cell, burning (oxidizing) essential biochemicals, efficiently destroying the cell.

Biologists know the evolutionary history of apoptosis and its origin with the mitochondria well enough, but the radical message has not been assimilated. The message is that the capacity, and the regulated mechanism, for self-destruction has been part of the eukaryotic life cycle since before the first eukaryotes. It is even older than the telomere counter described in the previous chapter.

Historically, apoptosis was discovered in mammals as part of the developmental process, and it is understood now as creative destruction in the interest of the whole organism. But apoptosis is also part of the body's death program. Later in life, the signals that trigger apoptosis are dialed up, and healthy, functional cells begin to duck out.

Hunger Games—How Apoptosis Has Been Reined In to Serve Larger Collectives

When you were an embryo in the womb, your hands began to grow as unshaped lumps of flesh. The fingers were formed first not as separate appendages but as an undifferentiated mass of cells. The five-fingered shape was formed when the cells in between the fingers destroyed themselves via apoptosis, much as a sculptor creating a human form from a block of stone would make fingers by chiseling out the stone in between them. Your brain, when developing in the uterus, sent out many exploratory neurons, and only some of them successfully connected to appropriate targets; the rest bowed out via apoptosis. So, too, your immune system relies on B cells that are precisely targeted to attack invaders. The way this is accomplished is via a huge variety of B cells being generated all the time, with all those that don't latch onto invaders committing suicide. Tadpoles maturing to frogs lose their tails through apoptosis. The many cells in your skin, your blood, and your liver that are constantly renewing themselves rely on apoptosis for recycling.

When cells in your body sense that they have become infected, they destroy themselves, hoping to take lots of the virus with them and thwart its further advance. Precancerous conditions can trigger apoptosis. A "famous" gene that controls this response is called *P53*, and every precancerous lesion on the way to becoming full-blown cancer must mutate to bypass the *P53* gene before it can become malignant.

It is easy for evolutionary biologists to understand these functions of apoptosis, because they sacrifice a small part of the organism for the benefit of the whole. Since every cell in the body carries the same genes, the community of interest is guaranteed. The ultimate mission of the body's somatic cells is to assure the success of the germ cells, which carry their DNA, in passing that DNA to the next generation. But the idea that single-celled yeast cells would commit suicide for one another's sake was considered heresy. Valter Longo discovered this phenomenon while still a grad student at UCLA in the 1990s. The discovery almost derailed his career, because so many in the community were convinced that what he reported was not possible. He persisted in the face of almost as many dismissals as Lynn Margulis had endured, at a similar stage in her career, when she put forward evidence for the origin of mitochondria from independent bacteria. It was through Longo's persistence and the clarity of his experimental design that he eventually won acceptance for his research and his result.

Yeasts are single-celled fungi. Colonies of yeast cells grow like crazy when there are nutrients available in their soup. They thrive in overripe fruit, helping to turn ripe apples into rubbery, prune-like husks. A yeast cell will "bud," creating little copies of itself that float off into the liquid, giving birth at rates as high as twice per hour. But when sugars in the environment are used up, the cells form spores, awaiting their next opportunity to thrive. Those that fail starve to death. What Longo discovered is that during food shortages, the cells don't wait for starvation—they jump the gun: they detect the food scarcity, and 95 percent of them sacrifice themselves via apoptosis. The yeasts chop themselves up, digest their proteins, and turn themselves into food for their cousins, allowing the remaining 5 percent a better shot at a new start on life, preserved as dried spores. The details that make this useful behavior possible are curious, because the 95 percent and the 5 percent are the same type of cells. Who or what is making the survival decisions? In the yeasts, what tells each individual cell whether to live or to die?

Perhaps there is some chance involved, a roll of the dice built into this genetically programmed behavior; or perhaps the cell colony as a whole is capable of detecting how many cells have sacrificed themselves and instruct the remaining ones to act accordingly, to become spores . . .

Here's why the neo-Darwinists were skeptical of Longo's results: imagine a mutation that would cause a yeast cell to hang back and wait for others to kill themselves. It needn't be absolute—even a small change of behavior that improved the odds for this particular cell from 5 percent to 6 percent would be sufficient. Natural selection would take care of the rest. Among the cells that formed spores, there would be more that carried this mutation. With each cycle of budding, starvation, and spore formation, there would be fewer of the 5 percenters and more of the 6 percenters, until their progeny would come to dominate the spore population. In time, another mutation would lead to 7 percent odds, and in this way, the adaptation that benefited the colony would slowly erode in favor of the self-serving behavior. This reasoning convinced biologists that single cells would never commit suicide for the sake of other cells, that this adaptation would never survive the neo-Darwinist scalpel.

In my view, there's nothing wrong with this reasoning. It is a plausible account of evolution and a reasonable expectation. But this should not be a basis for rejecting an experimental finding to the contrary. For one thing, experimental results are the reality check that keeps theorists honest. For another, evolutionary theory is not so well established that any predictions can be made with absolute certainty. The field is full of exceptions and mysteries, even more so than biology as a whole. What happened to Longo's paper was a serious error in judgment about what is known, what is plausible, and what is interesting and new. When the scientific establishment behaves in this way, progress is impeded.

Longo's results were controversial in the extreme. He is a thorough and meticulous experimentalist, but theorists took one look at his results and imagined that they knew better. This couldn't happen, they thought. Theory rules it out. You must have made a mistake. When Longo was finally able to get his work past peer review, there followed a barrage of attacks on his methodology and reasons why what he was seeing was not what it seemed to be. In all, it took about a decade for the reliability of his results to be established. And now, two decades on, the community has accepted the finding as though it were an isolated anomaly and so has failed to assimilate its deeper message.

For researchers who understand the evolutionary origin of apoptosis and its relationship to mitochondria, it should be no surprise that some ancient one-celled organisms already used this suicide mechanism. The same functions that now protect the body from infections once protected a yeast colony from starvation. A diseased cell already has a high probability of dying, so it is a small sacrifice for it to die promptly and voluntarily, for the sake of depriving the infective agent of cellular goodies. It is a larger sacrifice for a healthy, hungry cell among many other hungry cells to die for the sake of the others. It is tempting for the cell to wait and see if its neighbors will die to feed it.

Aging has this in common with cell suicide: the individual's elimination of itself benefits the health of the community. This is group selection in spades—not technically allowed in strict neo-Darwinism but apparently one of nature's most ancient and effective survival mechanisms.

How Apoptosis Contributes to Aging in You and Me

At the end of the previous chapter on cell senescence, we made a clear connection between shortening telomeres and whole-body aging. There is also a good case to be made that apoptosis contributes to whole-body aging in humans, though it is not quite so starkly clear. Cellular senescence in humans is wholly detrimental to us and has no function but to destroy us.* The role of apoptosis is more ambiguous. We need apoptosis to eliminate infected cells, defective cells, and cancer cells. It is very important that the body recognizes that these cells are causing trouble and that it gets them out of the way. So we cannot do without apoptosis at any stage of life.

Nevertheless, as we get older, apoptosis seems to be more and more on a hair trigger. Healthy, functional cells are eliminating themselves through apoptosis, and the whole body suffers as a result. Evidence for this comes from human pathology and also from one animal study in which mice were genetically engineered to have a slower, less sensitive trigger for apoptosis. Diseases of old age in which apoptosis is implicated include Alzheimer's, Parkinson's, ALS (or Lou Gehrig's disease), sarcopenia (muscle loss), osteoporosis (bone loss), and Huntington's disease.

After childhood, we are all losing brain cells much faster than new neurons are growing to replace them. In old age, we lose brain cells even faster. Alzheimer's disease (AD) is associated with this massive loss of brain cells. Could it be that relatively healthy neurons are simply eliminating themselves, and that this is a primary cause of Alzheimer's disease? There is direct biochemical evidence of this, but it is such a radical idea that most researchers continue to propose it cautiously as a hypothesis. Nonetheless, the gene variants that are associated with familial risk of AD all have to do with regulation of apoptosis.

* There is a popular theory that cellular senescence *must* serve a purpose for the individual, perhaps as a firewall against cancer. But the evidence is clear that short telomeres cause more cancer than they prevent. I described the situation in a 2013 journal article.

Another striking feature from the epidemiology of AD is the "use it or lose it" phenomenon. There is no particular physiological reason why neurons that are firing more often should be protected from apoptosis, and yet intellectual activity and late-life learning is well known to offer protection from dementia. The phenomenon seems similar to the beneficial effects in terms of health, longevity, and happiness to individuals in societies who have more family and friend connections. It seems almost as if nature is saying to us, "We know you're not strong enough to contribute to your community as a workhorse at this stage, but we still value your wisdom. If you're not exercising your brain, then maybe it's time to bow out." Could natural selection at the group level possibly be that smart? Certainly, within the framework of neo-Darwinism, this is inconceivable. But in an expanded view of how evolution works (multilevel selection), this kind of thing just might take place.

Muscle Loss, Parkinson's, and Menopause

Sarcopenia is medical terminology for the very familiar phenomenon that we lose muscle mass as we get older. It takes more work, more exercise, to maintain strength. Muscles eventually waste away and weaken despite the best regimens of exercise and nutrition. What causes the loss of muscle mass? Some of the reason is simply healthy cells eliminating themselves via apoptosis.

Almost everyone knows someone with Parkinson's disease. In its acute stage, it is an exquisite form of torture, making it harder for the patient to do anything at all, from walking to pouring a cup of tea. But we may not realize that early-stage Parkinson's is very, very common—perhaps universal in elderly people. Young people almost never get Parkinson's, and prevalence rises sharply with age. The cause of Parkinson's is the loss of a particular kind of nerve cell in a particular part of the brain (*dopaminergic* neurons in the *substantia nigra*). It seems that these cells are going AWOL, even though our ability to move depends on them.

One more example of apoptosis in action, making us old, is menopause in women. Women are born with many tens of thousands of eggs, but only a few hundred actually are promoted in the ovary and given an opportunity to be fertilized. Each month, there are dozens of eggs that ripen, but only one passes into the ampulla. The rest succumb to a process known as *atresia*, which is intimately related to apoptosis. The vast majority of a woman's eggs die unnecessarily, and then she runs out of eggs and fertility ends. This is a direct link between apoptosis and reproductive aging, which is the most relevant form of aging for evolution. In this view, a woman's reproductive "death" is a direct result of programmed cell suicide.

Alzheimer's disease, Parkinson's disease, sarcopenia, and loss of fertility afflict us all, to varying degrees, if we live long enough. All four are connected to programmed cell death, suicide genes in action. These suicide genes are the legacy of free radical–wielding mitochondria, whose capacity for creation and destruction has been absorbed into the modern eukaryotic cell.

The ancient function of cell suicide has resurfaced at a higher level, now co-opted for collective adaptations, including human aging. Aging in animals enforces a common, predictable life span, helping to prevent the dominance of any one individual or one gene type. Diversity is preserved for the health of the community.

Instant Replay

Apoptosis, or cell suicide, is the oldest form of programmed death, with a history of more than a billion years. Long before the first cells came together to form animals and plants, individual cells were able to detect when the population was too dense for the available food and to eliminate themselves for the good of the community. When a yeast colony is in trouble, most of the cells will die in self-sacrifice, digesting themselves, turning themselves into food for the remaining cells.

Like cell senescence, apoptosis also continues to exist to this day, and both these ancient modes of programmed death contribute to aging in humans. Replicative senescence and apoptosis are both part of the biochemistry that makes us old, a means to an end. As we get older, some of our healthy cells unaccountably commit suicide, causing atrophy of our muscles (sarcopenia) and loss of brain cells (related to dementia and Parkinson's disease).

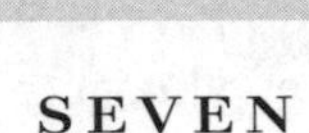

SEVEN

The Balance of Nature: Demographic Homeostasis

That which is not good for the beehive cannot be good for the bees.

—Marcus Aurelius

The Wide-Angle Lens

The main thesis of this book is that aging has been programmed into our genes by an evolutionary process, even though it's bad for our fitness as individuals. I hope that by this point in the narrative, you find this a little less strange than you may have at the beginning. We have seen reasons to believe that the neo-Darwinian framework with which most evolutionary biologists still understand their science is incomplete, but incomplete in what way? What is missing, and how can it account for the paradox that genes for aging have been fixed in the genome, despite the fact that these genes work against themselves, cutting off their own opportunity to reproduce? Aging has evolved despite the "fitness cost," despite the fact that aging and death cause an individual to leave fewer offspring behind.

There must be a benefit to aging. There must be a powerful something that more than compensates the negative of lost fitness for the individual. The benefit is certainly for the community and not the individual. And that benefit must involve some basic property of living things, since aging is found in so many diverse species. What is it?

My answer is that aging makes possible stable ecosystems by leveling out the death rate. Death at a predictable, programmed time prevents the tragedy of everyone dying at once—extinction in famines or epidemics. If evolution worked *only* at the level of individuals, it would be a dog-eat-dog world, rife with cannibalism and every form of vicious competition. Successful species would be successful at the expense of every other species, and, of course, that success would be temporary. We may compare it to a fly-by-night business that alienates all its customers and then must leave town hoping news of its malfeasance doesn't follow it; or to a parasite that kills all its hosts, growing quickly and very successfully until there is no one left to infect and the parasite dies out.

A predator must be mindful not to eat the last of its prey, lest it starve itself and (worse) starve its children. This point may seem obvious to you: evolution will not favor a species that devastates the other species on which its survival depends.

The Prudent Predator

A little less obvious is that the predator's best strategy is a rather extreme form of temperance. Suppose the predator were to get just a bit aggressive, and collectively all the predators eat half the available prey. They are rewarded by more energy to invest in making babies and feeding them. But look where their success leads! The next generation has twice as many predators and only half as many prey. The children will hunt desperately and drive the prey population down yet further, quickly courting a disaster.

In a famous and influential scholarly article called "How to Be a Predator," Stony Brook ecologist Larry Slobodkin put forward a demonstration of the predator's best long-term strategy. The best strategy, as it turns out, is to leave the prey population almost intact. The predator should take only the old who have already finished reproducing and the defective individuals to prune the herd.

This is the way to maximize the sustainable harvest and support the largest predator population in the long run.

This is an idea that has been explored and thoroughly accepted in the literature of ecology, and yet the idea that animals would actually evolve to pursue a strategy of restraint is considered anathema by most evolutionary biologists. Accepted principles of evolution predict that predators should evolve toward the strategy that is best for them *individually,* and that is always to consume all the prey available and to use the energy to reproduce as fast as possible. If this damages the prospects of other predators in the community, the result is a further benefit to the aggressive predator, whose progeny will come to dominate the community so much faster.

The species that abuses its ecosystem cuts off its own legs. Life is characterized by a tendency for exponential growth, a kind of upward curve that seems to explode out from nowhere. There is a lethal danger inherent in this kind of growth. It is easy for the population to become overextended. Maximizing this year's harvest can lead to many lean years in the future. We should not be surprised that life has evolved to defuse the population bomb.

But this idea has a surprisingly contentious history, and in fact, was at the center of the controversy in the 1960s that led to a mathematicians' coup d'état and an extreme proscription against the concept of group selection.

Why the Insistence on Pure Selfishness?

"Ecosystems don't evolve. Populations evolve, one gene at a time, within a fixed ecosystem." This has been an article of faith among evolutionary biologists for eighty years. But where does it come from? Is it a law of nature? A generalization from observations? A logical necessity? It's a

keystone in the foundation of neo-Darwinian thinking, and most evolutionary scientists will bristle if you question it. But it turns out that this is merely an assumption, an idea introduced by Ronald Fisher and his contemporaries to make their very complicated calculations a little more tractable.

Of course, there's nothing wrong with making simplifying assumptions. "Simplicity is the ultimate sophistication," Da Vinci quipped. Simplifications are at the heart of science and have led to amazing advances in physics. Einstein qualified, "Everything should be made as simple as possible, but not simpler."

It helps to recognize simplifying assumptions, to make them explicit. This is necessary so that later, if something goes wrong, steps can be retraced, and the culprit can be identified. The assumption that separate genes jockey for dominance within a static ecosystem has been part of evolutionary thinking for so long that scientists in the field have forgotten that it is an assumption at all.

Reductionism is the scientific practice of seeking to understand the whole by understanding the parts. For example, the chemist first conducts experiments on the behavior of a single molecule and then surmises the behavior of a gas made of very many independent molecules. Or the physicist first strives to understand the properties of an electron and then applies that knowledge to try to understand crystalline solids. So, too, in biology, we try to understand each individual animal and plant based on its chemistry and then use knowledge of individual behaviors to understand the behavior of an ecosystem.

But there is no guarantee that this process leads to understanding. The point about reductionism is that sometimes it works, and sometimes it doesn't. The kinetic theory of gases was enormously successful, based on modeling the behavior of many individual molecules bouncing around in an enclosure. But the properties of a single electron gave no hint about the field of solid state physics, which deals with electrons in crystals. In a solid, electrons don't behave like individual electrons at all but like instances of electron-stuff, linked inextricably to every other electron. There is one wave equation for all the electrons, and it is built into the laws of quantum mechanics that the electrons themselves cannot identify which individual they are but must respond as one connected whole.

So are individual animals and plants in an ecosystem like molecules in a gas, or are they like electrons in a solid? Are the needs and behaviors of a single bird sufficient to understand the dynamics of a murmuration of starlings, a bee sufficient to understand the hive, or is the whole greater than the parts in some essential way? There is no way to know in advance. Certainly the right approach is to try the simplest model first and assume that animals are independent. But if this hypothesis runs afoul of real observations and real ecosystems, then we'll have to go back and try a more complicated model.

Before mankind started wreaking our own havoc on the biosphere, it seemed to biologists that stable ecosystems were pretty much the rule everywhere. When Fisher and the neo-Darwinists sought to understand the mechanisms and rules by which evolution operates, they assumed a stable ecosystem as a backdrop. Genes come and go, while the ecosystem persists as a reliably steady environment. This simplifying assumption was logical enough. Indeed, the assumption was implicit, not even mentioned as such. Stability was assumed to be part of the way the world works and not worthy of notice.

Why Are There Stable Ecosystems?

Why do we see fields and forests, coral reefs and swamps that have a life of their own, a persistence longer than the individuals that live there? Traditionally, there have been two perspectives. One idea is that maybe it just happens that way. Maybe there is an "invisible hand" that keeps the populations in balance. Maybe each individual animal is fighting for its own kin alone, but the species is kept in check by the fact that other species are doing the same. The other idea is that there is active cooperation. Behaviors of individuals have been programmed in an evolutionary process that benefits not the individuals alone but the community and the entire ecosystem of which they partake.

In the first view, populations are governed by a law of supply and demand. When there is plenty of grass, the deer will flourish and produce more deer. When there are too many deer for the available grass, the deer will go hungry; some of them will die, and some will not have enough energy to lactate and feed their young.

This is the simplest hypothesis. Maybe there is no need for special "stability adaptations" that are coevolved across many species to permit them to thrive together in harmony. Maybe it just happens, like water seeking a common level, like a rock that comes to rest in the valley between two mountains. This sounds plausible enough, and it is the right place to start. It was the right place for Fisher to start when he laid the foundation for twentieth-century evolution. But it doesn't seem to be true, and much of the rest of this chapter will be devoted to showing you why. There are theoretical reasons: exponential growth is in the nature of biology, and exponential growth is hypersensitive and prone to overshoot. When computer scientists put together model ecosystems based on reasonable assumptions of individual behaviors, they find that populations tend to fluctuate wildly. And there are empirical reasons, as well: when an invasive species is introduced into a stable ecosystem, it is common to see the whole system implode, and a new equilibrium is established only much later. And year-to-year observations of local ecologies show that often a new variety doesn't establish itself by displacing the old variety one for one, but rather by reproducing explosively for a few generations and managing to survive the ensuing die-off.

The first view, the "invisible hand," has too often been an unquestioned assumption. There has been a prejudice against the second view, that whole ecosystem communities could have coevolved for their mutual benefit. There is a broad consensus that group selection represents a logical fallacy, and the kind of massive group selection creating a coadapted ecosystem is utterly implausible. We saw in chapter 3 where some of this prejudice comes from.

The Dogma

The neo-Darwinian idea of how evolution works sees individuals in all-out competition with one another, with the prize going to those who reproduce fastest. The notion of a communal "agreement" to restrain competition undermines this basic understanding. This is true even if the "agreement" we're talking about isn't communicated or recognized in the awareness but rather is cooperation that is coded in the genes.

The genes of animals and even plants that have no awareness can cause them to behave selfishly or cooperatively. In the neo-Darwinist picture, the fundamental relationship among individuals in a population is competition for genetic dominance, and restraint of reproduction in order to keep the population within limits would deny the core nature of neo-Darwinian competition. All appearances of cooperation, even of tempered ruthlessness, are illusions created by genes that look out for copies of themselves that exist in blood relatives.

In the book that was the clarion call of the mathematicians' hostile takeover, George Williams caricatured the claim of evolved population regulation and categorically denied its possibility.

"'[T]he population regulates its reproduction so as not to produce numbers in excess of what the environment can support.' Such expressions imply that the density regulation is an evolved adaptation of the population as a whole, and that without such adaptations there would be no numerical stability. These interpretations are utterly without justification."

Williams set his sights in particular on the work of V. C. Wynne-Edwards. It was clear to Wynne-Edwards that population control was ubiquitous in nature. He observed and documented dozens of examples in his articles ("The Control of Population Density Through Social Behaviour: A Hypothesis") and the book (*Animal Dispersion in Relation to Social Behaviour*) that was both the denouement of his lifework and the provenance of his downfall.

Populations can explode and overshoot their carrying capacity; whole species can crash to extinction, carrying their ecosystem with them into the dust. A stable ecosystem must not be taken for granted, and no species can afford to trash its ecosystem. Through historical examples, through studying the work of computational ecologists, and through my own computer modeling, I have come to believe that it is actually quite a trick to construct a stable ecosystem out of many unrelated species, each of which is trying to expand exponentially. I think the only reason that we see stable ecosystems in nature is that evolution has arranged it so.

Wynne-Edwards was onto something of central importance. I have come to believe that much of the harmony and cooperation that we see around us represents evolved adaptation and that this picture provides a natural background for the understanding of aging.

An ecosystem that is out of balance is in danger of collapsing to extinction, taking all its species out when it loses one or two that throw the populations out of balance. But a robust ecosystem is able to thrive and to expand its territorial dominion with all its constituent species moving and growing together. A robust ecosystem resists disturbances (from within or without) and is able to recover its balance when it is disturbed. A robust ecosystem persists over time by responding flexibly to changed environments and circumstances. Essential species within the ecosystem are protected and husbanded by a communal homeostasis.

The Rocky Mountain Locust: A Morality Tale

In the late nineteenth century, the American Midwest was plagued periodically by incursions of Rocky Mountain locusts. The appearance of this pest was devastating and unforgettable. Laura Ingalls Wilder wrote in her childhood memoir:

> *Huge brown grasshoppers were hitting the ground all around her, hitting her head and her face and her arms. They came thudding down like hail. The cloud was hailing grasshoppers. The cloud was grasshoppers. Their bodies hid the sun and made darkness. Their thin, large wings gleamed and glittered. The rasping, whirring of their wings filled the whole air and they hit the ground and the house with the noise of a hailstorm. Laura tried to beat them off. Their claws clung to her skin and her dress. They looked at her with bulging eyes, turning their heads this way and that. Mary ran screaming into the house. Grasshoppers covered the ground, there was not one bare bit to step on. Laura had to step on grasshoppers and they smashed squirming and slimy under her feet.*

In 1874, a swarm was described as being half a million square kilometers in area (for comparison, California is approximately 425,000 square kilometers). When a cloud descended, the land was denuded of everything green for many miles in all directions. The ground was thick with egg masses, ready to renew the plague the following year.

But the last reported sighting of a Rocky Mountain locust was in 1902. There are preserved specimens in museums and laboratories today, but no living locusts. Entomologists interested in the locusts' rise and fall travel to the glaciers of Wyoming, mining hundred-year-old ice for carcasses that they might study.

The Rocky Mountain locust drove itself to extinction by overshooting its sustainable population. These locusts did not die out because they were not individually "fit" in the sense of aggressive competition and prolific reproduction. Quite the opposite. They disappeared because they were too aggressive and too prolific. Individually, they were supercompetitors; collectively, they were a circular firing squad.

Where did they come from? Presumably some freak mutation created a monster, a species wildly out of control. Perhaps they derived from a European or Asian invasive locust. The mutant form was extremely mobile, depending on a hit-and-run lifestyle. They could descend on a landscape, destroy it, and move on. Swarms of locusts could fly hundreds of miles at a time to find new territories to pillage. Without such wings and such stamina, their heyday would certainly have been limited to a single season.

In one sense, the details of how they became extinct are a mystery. They were devastatingly efficient machines for turning leaves into eggs. Doesn't it seem curious that every last one of them could have disappeared? But on the other hand, their demise was completely predictable. Either they had to evolve into some less aggressive form, compatible with long-term survival of the plants on which they depended, or they had to disappear, victims of their own success. These forest ecosystems supporting insect life have survived for tens of thousands of years, built on a foundation many millions of years old. During that time, surely the Rocky Mountain locust was not the first incidence of the appearance of a superpredator. So we can take some measure of comfort in the assurance that ecosystems on this scale seem robust against invasion. Forests of the American Midwest grew back in the decades following the locusts' disappearance.

If this kind of event is rare in the present, it is perhaps only because natural selection has been at work for hundreds of millions of years, punishing species that were individually superperformers but were collectively unable to restrain their numbers and avoid population crashes.

Ecosystem Stability?

Ecosystems are so complicated that it is customary to begin study with artificially simple cases—for example, to try to understand one predator species and one prey species. Can the lion and the gazelle coexist in a hypothetical world where there are no other animals?

You might imagine something like this—suppose there are lots of gazelles and very few lions. Then competition among the lions is not an issue. The lions have an easy time finding food, and they can prosper and grow in numbers, restoring balance. On the other hand, suppose there are too many lions. Then there won't be enough gazelles to feed them all, and some of the lions will starve to death, which reduces the pressure on the gazelle population so they can grow back. This sounds like the makings of a stable system. It's called a "negative feedback system," because when it gets out of equilibrium in either direction, the tendency is for the system to move backward toward the equilibrium point.

But you might just as easily convince yourself of the opposite. If the number of gazelles is large compared to the number of lions, there will be more mother gazelles, producing a bumper crop of fawns that will barely be trimmed by the small herd of lions. The next generation will be even more overbalanced in favor of gazelles. On the other hand, suppose there are too many lions. The lion hunters in good times take only the old and weak gazelles, but when they are desperately hungry, they may feed on the vulnerable young. The result will be that the next generation of gazelles will be smaller than the present generation, while the large numbers of lion mothers will lead to yet larger numbers of lion babies. This is the opposite of negative feedback. It is an accelerating process, a positive feedback loop. The more there are, the more there will be. The runaway increase in the lion population will continue until the gazelles are all dead and gone.

We cannot decide from pure thought whether this highly simplified "ecosystem" is stable or unstable, and for complex, realistic ecosystems, this is all the more so.

An Ecosystem in a Bottle

When Leo Luckinbill was a grad student in the Zoology Department at UCLA in the late 1960s, the computer was a new toy that allowed ecologists to solve equations for population growth for the first time. The field had been developed by mathematicians and physicists. Their *modus operandi* was the most fruitful methodology of classical physics. First, they would abstract from a system an equation that describes the way in which it changes over time given its present status. There was already a rich theoretical literature of such differential equations in ecology but very little contact with experiment. "As a consequence, mathematical models of simple predator-prey systems have been developed to a degree of sophistication far advanced beyond the support of empirical experimentation," Luckinbill noted. The plan for his dissertation was to compare the behavior of the simplest solutions to that of the simplest ecosystem he could create in the lab.

Luckinbill prepared a water bottle with appropriate nutrients and seeded it with two protist species, one of which is known to eat the other. He chose microbes because they were easier to work with than lions and gazelles. The convenient forty-eight-hour life spans of the protists he used would permit him to finish his dissertation before he died of old age. But he hoped and expected that the results would offer a broader lesson in ecology that could be applied to the macroscopic world.

In ponds the world over, *Didinium* thrives on *Paramecium*, and Luckinbill thought it would be an easy matter to establish this simplest ecosystem in his lab by providing food for the paramecia and allowing these ciliates to provide food for didinia. He would observe the system through a microscope and track the numbers of the two species as they changed over time.

The result of the experiment was more interesting than he had anticipated. No matter how he tried to set up the advantage of one species or the other, Luckinbill always found that one of two things would happen: either the paramecia would fend off the predators and all the didinia would die out; or else the paramecia would be wiped out entirely by the predators, and then the didinia would starve to death. This would happen quickly, within the first three days. He was unable to establish a stable community with predator and prey in balance, so long as they were freely mixed in one bottle.

Luckinbill's careful observations and analysis helped launch a new wave of thinking in population ecology. The differential equation that theorists had used to describe this situation is "neutral stable." It has stable solutions and unstable solutions, either of which can cycle over time. Theorists had imagined that nature had found ways to keep the stable solutions, but in lab experiments, stability seemed to be elusive in a way the equations did not predict.

Where had the theory gone wrong? The equations were built on the assumption of immediate feedback. The way in which the equations worked had no time delay built in for growth and immaturity. This works fine so long as generation time is short, so population turnover is quick and population growth is slow. But what Luckinbill found was that populations could change rapidly in practice. If mortality was low, population could double in a single generation, and the equations weren't built to handle this. The system or predator and prey had a propensity to overshoot that the equations did not account for.

Real ponds are different from the ecosystem-in-a-bottle scenario because there are many species that hold each other in check. The didinia in the bottle had no predators and could die only of starvation, but in a real pond, there are many insects eager to make a meal of didinia. It is also important that the pond is large and imperfectly mixed. On any given day, the paramecia may be wiped out in one area while thriving in another. The didinia can chase the paramecia around the pond, ever migrating toward fertile hunting grounds. On a larger scale, the world's great plains and forests follow this same pattern, cycling from decade to decade as predators move in, clean up, and then leave the area to recover over several years.

It's All in the Timing

We think of the self-regulating marketplace as a model of a naturally stable system based on negative feedback. The law of supply and demand operates by automatically lowering the price of goods that are in oversupply and raising the compensation for goods (or professions) when there is a shortage. But suppose this feedback operated with a delay. For example, there may be a doctor shortage driving up medical fees and salaries, but the only way to get more doctors into the system is to send students through four years of medical school and three more years of internship and residency. Worse, medical schools may already be operating at capacity. Getting each new medical school up and running might require a decade. And the doctor shortage is now! In our economic world, there is communication and planning, and people operate with some amount of foresight, so this hypothetical situation is less likely to arise. But as an example of how nature operates, it can offer some insights. If people built medical schools only in response to present shortages of medical personnel, we might have large numbers of people dying for lack of hospital care before a new supply of doctors arrived on the market. By the time the situation began to correct itself, there would be more medical schools than the country could support, training more doctors than the community ultimately needed. After a decade or more of schooling and residencies, a doctor is highly motivated to continue in his profession for the full duration of his career. So the dearth of doctors might be followed by a glut of doctors for a generation to come. The profession would lose prestige and the ability to command a professional salary. People would have to have fanatical devotion to begin medical school under such circumstances, and so the stage would be set for an even steeper shortage of doctors beginning thirty or so years after the doctor glut.

Without adaptations that specifically address stability, ecosystems could well behave in this way. The negative feedback in ecosystems comes with an insupportable delay. Larger animals, especially, have long generation times. They cannot afford to reproduce based on the present availability of food, because the next generation is likely to starve. If caribou living in the tundra were to eat all the ground cover that was available to them, their offspring would starve, and the tundra might take decades to regenerate itself. The caribou in Alaska are smart enough (this intelligence is in their genes, not their brains) to eat lightly and to mate only in alternate years. Their cousins (same species) that live farther south can afford the luxury of reproducing every year, and so they do. This is an example of cooperation that avoids ecosystem collapse—individuals behaving in a way that is good for the species but bad for their own individual fitness.

Crash of an Arctic Hunting Preserve

Barren and mountainous, Saint Matthew Island in the Bering Sea supports low-growing tundra in the summer and not much to eat in the winter. In 1944, the largest animals inhabiting the island were voles (mouselike rodents in the same family as lemmings and muskrats) and a few foxes that chased them. Thinking to create a hunting preserve, wildlife managers introduced twenty-nine reindeer that year, imported from the island of Nunivak, two hundred miles to the east.

The reindeer flourished, their population growing by about a third from each season to the next. That may sound like an extraordinary rate, but it represents less than one surviving fawn per year for each pair of parents. The ability of the population to expand rapidly is beneficially adaptive in an empty niche and may be a lifesaver after a natural disaster. So the reindeer population followed a trajectory typical of an exotic species that is successfully introduced: exponential growth. Naturalists estimate the carrying capacity of the island at about two thousand reindeer, and the population crossed that threshold around 1960.

Such is the relentless logic of exponential growth that just four years later, the population was six thousand reindeer. The winter of 1964 was severe—not a dramatic departure from what the reindeer expected, but more snow than usual. By the end of the winter, nearly the entire population had starved to death. An expedition the following year counted forty-two stragglers (and shot ten of them in the name of sport and science). Reindeer live typically eighteen to twenty-two years, so the entire saga—boom to bust—had unfolded within the lifetime of a single reindeer.

Scientists are privileged to witness dramatic die-offs like this only rarely. The cases that have been documented all involve either the influence of encroaching civilization or animals that are newly introduced into a favorable, virgin habitat. Of course, this is to be expected: in a natural habitat, the extinction event would most likely have occurred long ago, before humans were around to record it. Still, we have to wonder, when populations undergo dramatic local extinctions, do they vanish without a trace of a legacy, or do some animals in some places survive? Do the survivors tend to have a different genetic makeup than those who succumbed? And does the event leave its imprint on the genome?

Yes, of course this is what we should expect. This kind of account sounds to you and me like classical Darwinian logic. The population that learns to detect the limits to growth early avoids a crash; the population that doesn't succumbs to extinction. Over time, those populations that carry genes that help them to moderate population growth in advance of food shortages persist, and their legacy survives, while competing populations may vanish.

The surprise is not that this kind of dynamic should occur in nature but how vigorously this story is denied by mainstream evolutionary theorists. A popular ecology text says: "There are powerful and fundamental reasons for rejecting this 'group selectionist' explanation . . ." How did neo-Darwinists come to think this way? And how could they cling so insistently to theory in the face of countervailing evidence?

Neo-Darwinian theory is based on the assumption that genes can spread in a few individual lifetimes, and population sizes can be counted on to remain stable during this time. But we just saw that it took only four years for the reindeer on Saint Matthew Island to grow from two thousand to six thousand. The disastrous population crash that followed brought the reindeer to the brink of extinction, and it was a direct result of their prodigious "success" in populating the island.

These reindeer came originally from Nunivak Island. Why didn't they overgrow there? I have no certainty to offer, but I suspect it had to do with the wolves and other predators that were part of the Nunivak ecology but missing on Saint Matthew. Their reproduction was attuned to a higher mortality rate, and in the context of the Nunivak ecology, they were able to maintain a stable population. Freed from the predators that killed them, they likely grew in numbers until they starved themselves to death.

Evolution of the Predator to Preserve the Prey Population

The idea that Larry Slobodkin articulated has been called the "prudent predator hypothesis." Predators develop restraint. Over evolutionary time, they learn to trim the prey population but not to cut too deeply into their numbers. Of course it is not the brains but the genes that are doing the learning, and they learn by trial and error, repeated extinction of those that are too greedy. Evolution has taught them to leave a healthy prey population for their children's benefit.

This process has never been observed by evolutionary scientists because evolution takes a long time. But a closely related phenomenon is well known in the field of epidemiology. It is common for deadly bacterial parasites to evolve toward being less deadly. Those that kill their hosts too quickly are less likely to be passed on, compared to those that allow the hosts to survive a long time while contagious. Because bacteria have a short generation time, and because medical science is always on the lookout for the next epidemic, this evolution of infectious bacteria from a more virulent strain to a more benign form has been observed often.

A Lone Mathematician Supports the Theory of Cooperative Evolution

The direction of evolutionary biology in the twentieth century was centrally influenced by interlopers from mathematics and physics. Alfred Lotka and R. A. Fisher were mathematicians. Max Delbrück and Leó Szilárd were physicists. George Price and J. B. S. Haldane were physical chemists. Even Erwin Schrödinger, a famous father of quantum mechanics, wrote a seminal book *What Is Life?*

All these men embraced the neo-Darwinist model of gene frequency in a static population as their primary model of how evolution works, and together they did much to establish this paradigm as the standard way to view natural selection.

Standing alone, an academic cry in the wilderness, was the mathematician turned ecologist Michael Gilpin. Gilpin was familiar with the use of computers in physics to help understand systems that are too complicated for equations that can be solved by hand. He saw simple, paper-and-pencil equations used to justify the conclusion that evolution could only produce selfish genes, and he had a hunch that more complex equations, tractable only in a computer, might produce a different picture. He created a computer simulation, forerunner of the virtual reality games that have become popular today. In 1975, this was pioneering work, the first time that evolution and ecology had been integrated into a computational system.

Gilpin had just received his Ph.D. in mathematics from Stanford, and, around the time of the first Earth Day, he developed a keen interest in ecology. He determined to set Wynne-Edwards's thesis about natural population regulation on a firm mathematical foundation. Computer resources were cumbersome and expensive at the time, requiring great patience and significant expertise specific to the machine itself in order to perform complex calculations. But in this area, Gilpin was right at home. He had the mathematical chops to argue his point with great clarity and precision.

Gilpin wrote a closely argued monograph on the subject of predator restraint. He assumed that predators that ate more would be able to support a higher fertility rate and would thus come to dominate their local population. However, if they became too greedy in the aggregate, their children (and the children of those around them) would all starve in short order. He demonstrated with computational models that famine and extinction could be powerful evolutionary forces at the group level. Interspersed with his computer calculations, he included logical and mathematical arguments to show that the assumptions behind his calculations were reasonable and that the real world was likely to be, in every way, more conducive to the evolution of predator restraint than his model.

Gilpin showed that the balance of evolutionary forces could be expected to create predators that were restrained and conservative, holding back to protect their prey for future generations. The results were not perfectly prudent like the long-term optimization visualized by Slobodkin, and they were not voraciously selfish as predicted by the equations of neo-Darwinism. They were a compromise between shortsighted greed and long-term prudence, between perfect selfishness and perfect altruism. Gilpin's monograph attracted scorn and some shallow criticism, but for the most part, it was ignored.

Gilpin showed that famines are a powerful evolutionary force, because they quickly and efficiently eliminate populations that have evolved to be too aggressive. Evolutionary theorists in the 1960s and '70s had wondered what possible group selection mechanism could be efficient enough to counterbalance the directness and efficiency of individual selection for higher rates of reproduction, and here was their answer. Populations can wink out of existence in a single generation when their food supplies become exhausted. Predators must "learn" to limit their reproduction in response to the availability of food. Halting the growth of a predator population is like stopping the course of an ocean liner; there is considerable inertia in the exponential expansion, so growth will only be halted in time if restraint is begun *before* food becomes scarce.

Gilpin's work is badly out of fashion today, and the man himself has moved from a career in mathematical ecology to the more urgent business of conservation advocacy in his adopted state of Montana. When the pendulum swings back and the importance of evolved eco-dynamics is recognized, Gilpin's pioneering computer models may someday be honored as an insight ahead of its time.

Population Regulation as an Evolutionary Force

Stable ecosystems don't come for free. If we find homeostasis in nature, this is likely to be the result of a long process of coevolution.

To a nonspecialist, even to a scientist who has not been specially trained in population genetics, this may sound like common sense. But it is a process outside the realm of neo-Darwinian thinking. It is a good candidate for a way to extend classical evolutionary theory beyond the selfish gene. Population stability is essentially a collective attribute; it makes no sense to talk about an individual being ecologically stable or unstable. And crucially, population instability as an evolutionary force can act very fast and with devastating results.

The essential argument that cast group selection in a bad light beginning in the 1960s was that it takes too long to work; group selection requires whole populations to go extinct, and this is a very infrequent event compared to individual deaths. It was argued that any trait that offered a benefit to the population at the expense of the individual would be extinguished by individual selection within the group in just a few generations, long before its benefit for the group could become apparent. We now see that a few generations is plenty of time for a population to outgrow its resource base and crash back to oblivion.

George Williams, John Maynard Smith, and other astute scientists who made the argument for individual selection noticed that population extinctions were relatively rare, and they accepted that this is the way nature works without ever asking why. But by Gilpin's logic, population extinctions must have been much more frequent in earlier evolutionary times, and the ecological stability we see today is a feature of life that evolved so long ago that we are now tempted to take it for granted. All life is accommodated to an ecosystem, and cycles of birth and death are adapted to prevent the worst kinds of abuse that might bring an ecosystem crashing down. Williams and Maynard Smith were assuming the condition they had set out to prove.

Several academic papers written in the 2000s have surveyed wild ecosystems and concluded that sometimes they can change very rapidly. In less time than the life of a single individual, the ecological context can change dramatically. This finding removes the foundation from the claim that group selection is much slower than individual selection.

Living Lightly on the Earth

The heart of Earth is green; wherever you look in nature, food chains are bottom-heavy. There are enormous numbers of photosynthetic bacteria, algae, and green plants, smaller but still prodigious populations of insects that feed on them, much smaller populations of birds and small mammals that eat insects, and tiny populations of top predators. But a bottom-heavy food pyramid could never have arisen through a neo-Darwinian process of intense individual competition. Why? Because intense competition for resources would mean that at each trophic level, the prey species would be nearly wiped out by the predator species before the predator species began to experience the limits to growth. The earth would not be green but brown, because every leaf and every blade of grass would be scarfed up before it could mature. In order to have as many children as possible, I should be packing away biomass, eating as much as I can eat, grabbing more and more of the communal pool of food for me, me, me. According to classical neo-Darwinism, I should continue hogging food even after I'm no longer hungry, just to deprive my neighbors of food, so as to assure my relative competitive advantage in seeding the next generation. (The technical term for such behaviors among evolutionary theorists is "spite," and it's not hard to see why.) The collective result of many individuals behaving selfishly is a tragedy of the commons in which the shared food supply is overexploited, and the whole community loses.

While we occasionally observe some of this in nature, the remarkable thing is how much of nature has managed to avoid this fate. What we generally see is that at each trophic level, the predators are sitting lightly on the prey population, skimming off the excess, but permitting the community of prey to thrive close to its maximal extent. Moderation in all things, "nothing in excess"—an ancient admonition inscribed at the Temple of Apollo at Mount Parnassus in Greece.

This is remarkably close to Slobodkin's calculation of "How to Be a Predator." Resource management is another consequence of exponential growth: the larger the prey population, the more biomass it is generating, the more that can be sustainably skimmed for the predator's dining pleasure. Maximizing the size of the pool of prey is the best strategy for sustainably supporting a large predator population.

But from the point of view of each individual predator, the prey population is a huge, wasted resource. Any individual predator that grabs more food has a chance to turn the excess into babies so that her progeny will take over the population, and every one will then inherit the parent's selfish propensities. If you think like a neo-Darwinist, this must be your prediction: that runaway selfishness will lead to a tragedy of the commons, and all will starve.

How have so many species in nature managed to avoid this fate? We may imagine many, many population-level die-offs and species-wide extinctions over a long period of time before genes "learn" to behave less selfishly. In most neighborhoods, predators behaved selfishly and their children starved; in a few, by chance, a different arrangement of genes may have led to predator restraint. It is those few that survived, and when neighboring areas began to recover from the devastation, those few were poised to migrate in and to spread.

Amid many local extinctions, sustainable ecosystems persist; they are the survivors. The most robust ecosystems often enough include animals subject to aging and other built-in systems that moderate population growth in the best of times, avoiding overshoot that courts disaster. (Territoriality and density-sensing moderation of reproduction are other adaptations for population stability.)

Of course, this is a description of "group selection" in spades, and the neo-Darwinian consensus is that evolution can't work in this way. But it is also the only explanation for the persistence of the green earth that we see around us.

Instant Replay

The evolution of population regulation has a long and contentious history. The idea that animals could evolve an ability to sense population density and hold back from reproducing too rapidly for the sake of avoiding a population crash later on has been repeatedly subject to scorn and ridicule. In the 1950s, V. C. Wynne-Edwards documented this topic extensively with his own observations of nature, but he was told by mathematical theorists that he was misinterpreting what he saw, and his stock dropped precipitously at the pinnacle of his career.

The central question is whether ecosystems are able to maintain their balance as a result of an "invisible hand," or whether the constituent species, and ultimately the individuals, have evolved to sacrifice some of their individual fitness for the sake of coexistence. Since the repudiation of Wynne-Edwards's book in the 1960s, the mainstream of evolutionary scientists has lined up behind the "invisible hand."

However, we have seen in this chapter a variety of evidence that ecological stability does not come for free and must be an evolved attribute. We saw how, in a simple laboratory demonstration, predators and prey cannot survive for more than a few generations before one drives the other to extinction. In microbes, we were reminded of how evolution moves toward decreased virulence, which has been frequently observed and is well known. Since predators are functionally just larger parasites, this lends credibility to the idea that unselfish behavior can evolve for the sake of the next generation.

There is also theoretical support for this idea. In the early computer models of Michael Gilpin and in many models since (including my own), the interaction of evolution with ecology has been demonstrated, and the result is that a compromise emerges—behavior that is selfishness tempered by moderation. In other computer models that simulate populations in a complex ecology based on purely selfish behaviors, the population numbers are observed to fluctuate wildly. This is so different from what is observed in the real world of animals and plants that we must conclude that limits to selfishness have managed to evolve.

EIGHT

So We All Don't Die at Once: Wiles of the Black Queen

We are here to help each other get through this thing.

—MARK VONNEGUT, RESPONDING TO HIS FATHER KURT'S QUESTION, "WHY ARE WE HERE?"

So We Don't All Die at Once

And so we come to the core of this book, what I believe is the evolutionary force that has led to some form of aging in most animal species. Having come this far, you won't be surprised that it has to do with keeping populations from growing too fast, only to collapse and risk extinction.

In animal populations, death from external causes tends to clump together in time. Think famine. Think epidemic. Think droughts and storms and natural disasters. If there were no such thing as "old age," populations would just keep growing, making hay while the sun shines, until the crowding began to cause food shortages, and weakened animals in crowded conditions could lead to epidemic outbreaks. We saw in the previous chapter that once food becomes scarce enough to limit reproduction, it's probably too late to avoid full ecosystem collapse. Without aging, animals would only be dying in famines and epidemics, and then they would all die at once.

It stands to reason that local extinctions like this are a powerful Darwinian force and that we ought to expect that populations are broadly adapted to avoid them. But this whole way of thinking lies outside the realm of evolutionary dynamics encompassed by the mainstream of evolutionary research. Believe it or not, avoidance of extinction is not considered a substantial target of natural selection. Neo-Darwinists argue that genomes cannot "think ahead" to avoid extinction, and do not recognize the possibility that repeated local extinctions may have selected combinations that are less prone to population collapse.

You may think that population regulation comes for free, that it is in the nature of nature that when there are too many rabbits, then some of them will starve and the next generation will be smaller. But population overshoot and mass die-offs are also in the nature of nature, as we saw in the previous chapter. We die one by one of internal causes, and thus we avoid the external calamities that kill us all at the same time.

Aging has the effect of leveling the death rate in good times and bad. Aging makes possible the relative stability of ecosystems with populations that fluctuate within a range but don't soar past sustainability only to crash and burn. Aging evolved because populations of animals that don't age tend to grow too big for their food supply, followed by extinction. Unregulated, ecosystems tend to be "chaotic." Your common-sense interpretation of this word is right on the money, but this word has a mathematical meaning, which also applies, as we shall see. It is to stabilize populations, to avoid the chaotic cycles of boom and bust, that aging has been selected and has spread through the biosphere. This is a statement of the Demographic Theory of Aging, which I have introduced into the evolutionary literature.

The traditional objection to all theories that say aging is preferred by natural selection is that aging must evolve "uphill" against a powerful selective current. Selection against aging is direct and immediate. Selection for aging is long term and indirect. But the Demographic Theory overcomes this objection, because extinction can act with great rapidity, wiping out an entire population in a single lifetime. (We saw as much in the example of the Saint Matthew reindeer in the previous chapter.) The Demographic Theory is a solution to the century-old conundrum, how did aging evolve? And it is an answer to the neo-Darwinists' credendum that natural selection works on individuals more efficiently than on groups.

What Doesn't Kill Me Makes Me Stronger

Recall from chapter 4 that when animals are subject to hardship, paradoxically, they often live longer. Starvation, exertion, punishing heat or cold, small amounts of poison, or even radiation can lead to life extension.

"Hormesis" is the technical word for this strange phenomenon of nature: overcompensation in adapting to stress. We expect undercompensation, because nature is economical. Where we find overcompensation, we scratch our heads—we know something funny is going on.

For example, in the winter, it might be 30° Fahrenheit outside and you keep the temperature in your living space at 65°, because that's the lower end of your comfort zone, and you don't want to waste energy heating your house warmer than is necessary. In the summer, the temperature outside might be 90°, and you might air-condition the house at 75°, because that's the upper end of your comfort zone. It would be a strange thing to set your thermostat at 65° in the summer and 75° in the winter. You have better things to do with your money than to throw it away on heat or air-conditioning that is more than you need.

When an animal is unthreatened by bacteria or parasites, when life is easy, plenty of food, little exertion, a warm, clean environment, you'd expect it to do well and live a long time. Instead, we find that it does well and lives a short time. Life span is shortest when conditions are ideal!

We expect that when there are no challenges, the body's defenses are down. Perhaps the immune system takes a vacation when there are no invasive microbes, the muscles weaken when there's little work to do, and the body runs in a mode that's less fuel efficient when there's plenty to eat. Then, in the presence of infectious bacteria, we'd expect the body to adapt and the immune system to become stronger, so we expect it to do almost as well with the bacteria as without them. But why would we do better? Why would the body overcompensate for the stress of infection?

Stated the other way around, we know the body is capable of mounting a heroic response to stress. We understand that it would relax when stress is absent. But why would it ever let its defenses down so much that life is actually shortened in the absence of stress?

The word "hormesis" refers to a whole class of adaptive responses in which the body is programmed to live longer, to dial down aging and keep the body stronger for a longer period of time under various kinds of stress. There are two things we can say about hormesis before we look at the details. First, if the body does better under stress, it means that health is being held back when it is not under stress. In other words, living longer under stress is only possible if evolution has left some room to maneuver. The life span must be evolutionarily programmed to be shorter and the body less strong when life is easy. Some strength and longevity is kept in reserve to be applied only in times of stress. Second, hormesis helps to level out the death rate and soften the impact of famine and hard times. Just when there is a high death rate from predators or famine or a natural disaster, aging backs off and takes a smaller toll. The net effect is to reduce the risk of overpopulation in times of plenty and also to help safeguard the population with extra strength and longevity when the risk of extinction is high.

Hormesis is built into the genealogy of aging at a deep level, and this should be a clue about how aging evolved and what purpose it serves. In fact, the oldest and best-preserved genes connected to aging mediate the adaptive response to stress. There are genes that control hormesis that we share with our ancestors going back to the earliest protists. Two of the best-known examples concern starvation and physical exertion. The first is the insulin pathway. Insulin in our blood signals an abundance of sugar. But long before blood was invented, insulin was produced in response to satiety, and insulin hastens aging, shortening life span when there is plenty to eat. The second is the ROS metabolism (reactive oxygen species, a.k.a. free radicals). Energy-intensive activity is often a response to stress. When we shiver to keep warm or run to escape a hungry tiger, the extra energy production generates ROS, and the ROS tell the body to go into a stress-resistant mode. The signal is actually a surrogate for the probability that my sisters and cousins are probably being killed by the stress, and it would be a good time for me to not get old and die, for the continuity of our community. We noted in chapter 1 that despite the fact that oxidative damage comes with aging, attenuating oxidative damage with antioxidants actually shortens life span.

Now we are in a position to understand the reason for this paradoxical behavior. Hormesis helps to level out the death rate and stabilize populations. Natural selection has arranged to adjust death from old age so that it complements the natural cycles of boom and bust. When many individuals are already dying of starvation, death from aging takes a backseat. But when there is plenty to eat, death from old age takes its largest toll.

Paraquat

Paraquat is a defoliant that burns everything it touches. It was used in the drug wars by American planes spraying marijuana fields in Mexico, until word got out that too many human bystanders were dying.

In the McGill University laboratory of Siegfried Hekimi, life span of roundworms is extended remarkably by adding paraquat to the medium in which they swim. Low doses of paraquat have little effect, and high doses kill the worms, but if the dose is adjusted to an optimum, the worms live 70 percent longer.

Why does it work? Hekimi does not think in terms of population regulation, and "hormesis" is a word he avoids. He prefers explanations in terms of the proximate biochemistry. So-called reactive oxygen species, or ROS, more commonly known as "free radicals," indeed do damage, but these powerful oxidizing agents are *also* the signal that turns on the body's defenses. Small amounts of paraquat extend life span in worms because they mimic the signals that activate powerful metabolic defenses.

But why stop there? We can ask why the body doesn't turn on those same defenses when it is not being poisoned. And we can ask how these corrosive chemicals came to be a signal for something life preserving.

My answer to the first question is that the body has been programmed to keep some fitness in reserve for tough times, thereby helping to soften the impact of those big stress events that threaten

extinction. And to the second, I answer that ROS are generated during physical exertion, which is a sign of stress. Long, long ago, our ancestors the worms learned to distinguish a stressful environment from a cushy existence by the presence of ROS, and the biochemistry of that stress signal has been preserved and passed down the evolutionary tree for half a billion years.

Mathematical Chaos and Ecological Chaos

In 2006, I first introduced the Demographic Theory of Aging in an article in *Evolutionary Ecology Research*. The article included a proof that if you think like a neo-Darwinist, with selfish genes only subject to natural selection, the result is inevitable population collapse. With the same math tools, I also demonstrated that aging can rescue the population from this fate. From chapter 2, you know how wary I am of mathematical proofs in biology. So this one comes with a warning in advance: think for yourself about what it demonstrates and what it misses and what the unstated assumptions and hidden loopholes are.

The proof is framed in terms of the simplest ecosystem, with two species, one of which lives by eating the other. You can think of them as predators and prey, rabbits and grass. Ask about the growth rate of the rabbit population and the growth of the grass—which grows faster: the rabbits or the grass?

Both rabbits and grass are evolving, and natural selection will reward the individuals that reproduce faster. But the grass has long ago reached a fixed limit. Grass gets its energy to grow from the sun, and it takes a month for a blade of grass to accumulate enough energy to make another blade. Photosynthesis is about as efficient as it can be. It is much more efficient than any solar panel that humans have built so far, and so the grass has reached a ceiling of reproductive speed.

The rabbits, however, can evolve to eat more grass, and the more they consume, the more energy they have, the faster they can reproduce.

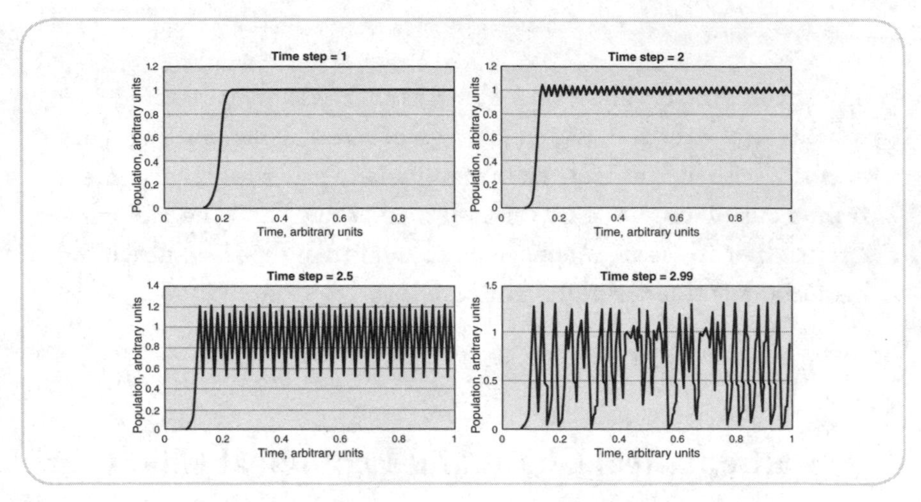

Fig. 2. The fastest-breeding rabbits take a one-way trip to extinction.

Suppose one individual rabbit evolved to forage more aggressively than her sisters, to eat the grass down to the roots. She would have the energy to produce more baby rabbits, who would inherit her instinct of aggressive foraging. Her progeny would soon take over the hutch. But the victory would be short lived. When grass is cut to the nub, there is no more green blade to absorb sunlight, and it takes a long time for the grass to recover. All those extra babies that provided a selective advantage for the aggressive forager at the individual level would starve to death in the next generation.

Rabbits have evolved a faster doubling time than grass, and in theory, the rabbits could evolve to reproduce faster yet by eating more grass and having more babies. In fact, in the past, rabbits probably did evolve to reproduce faster and more efficiently than at present, but the rabbit populations that did so suffered rapid demise. This is according to a simple and surprising piece of mathematics that was developed in the last eighty years. Here is the solution to the equation for logistic chaos, discovered in 1973 by Mitchell Feigenbaum. Think of the graphs above as the way the population changes over time.

In the 1970s, a decade before the personal computer, came the programmable calculator. They couldn't host video games or even support a word processor, but with ridiculously painstaking precision, you could tell them to repeat a series of arithmetic operations in a specific order. For numbers geeks like Mitchell Feigenbaum (and me), they were a dream toy and could provide hour after hour of amusement.

The calculations that demonstrate chaos are simple but tedious. No one before Feigenbaum had imagined that such simple equations could produce such intricate complexity.

1. If the rabbits reproduce more slowly than the grass, then their population will find its carrying capacity with no overshoot whatsoever. The rabbit population will be absolutely stable.
2. If the rabbits evolve a faster reproduction rate, their population will tend to overshoot and then oscillate around the carrying capacity.
3. If the rabbits evolve to reproduce 2.5 times as fast as the grass, the population will start to oscillate significantly.
4. At 2.9 times as fast, the oscillations are severe and irregular, and the population is in trouble.

And—though it is not depicted in the figure—should the rabbits ever evolve to reproduce three times (or more) faster than the grass, their population will overshoot too far above carrying capacity and crash down to extinction after the very first cycle. Of course, the number 3 comes from mathematics and not biology. There is no biological reason the rabbits couldn't reproduce three times as fast as the grass (and in some environments, the rabbits' speed advantage is already more than three).

What this means is that evolution has had to find just the right rate of reproduction, taking into account the local sun and rain that controls how fast vegetation can grow, and taking predators into account, as well. The rabbits that reproduce too slowly will be outcompeted by rabbits that can breed faster or with larger litters. But any rabbit that carries this game too far risks dooming its grandchildren to starvation.

Time Out for Definitions

It may not be obvious that it is even a meaningful question. How can we define growth rate in a way that makes sense for such different living things?

It's approximately true that in the absence of competition, life expands exponentially. When the grass is thin, it takes perhaps a month for it to grow twice as thick. If we wait another month, there is four times as much grass, and so forth.

Rabbits follow a similar mathematical law. If we start with one hundred rabbits of assorted ages, it may take a month before we have two hundred rabbits and another month before we have four hundred rabbits.

So we can measure the speed of reproduction as the time it takes to double the quantity of grass or double the number of rabbits.

Population Regulation Becomes a Universal Principle for All Animals

All animal life ultimately depends on plant growth for sustenance. Even the top predator cannot grow faster than the food-producing plants that support the whole community.

Plants have nothing to worry about, but animals cannot afford to reproduce more than about two times faster than the plants at the base of the food chain. From the perspective of simple mathematics, this seems obvious and straightforward, but it directly contradicts a fundamental principle of neo-Darwinism, which says that animals evolve to reproduce as fast as physically possible.

So here is a powerful Darwinian force based on group dynamics that is in direct opposition to the individual force of natural selection, always pushing for more efficient growth and faster reproduction. The whole life history, from birth to maturity to reproduction and death, must be shaped so as to be quick and efficient, but not too quick and efficient.

This principle changes the context for evolution of aging. It is always assumed, based on what is good for the individual but not for the group, that a longer life span offers an evolutionary advantage and that natural selection will always tend to increase life span, all things being equal. This is not true. Natural selection must adjust the fertility and the life span to create a net speed of reproduction that is matched to the ecosystem—not too high or too low.

In 2006, I demonstrated that either lower fertility or shorter life span can evolve to avoid population chaos. I believe that nature has found both these solutions, but there are reasons that natural selection prefers higher fertility and shorter life span to the other way around, as we'll see later in this chapter.

The Rabbits in Australia: A Cautionary Tale

The continent of Australia has evolved its own species in isolation from the mainland. The only mammals that are native to Australia are marsupials, like kangaroos and koalas. In 1859, William Austin came from England to visit his uncle, and he anticipated loneliness far from home. How might he continue his favorite sport? "The introduction of a few rabbits could do little harm and might provide a touch of home, in addition to a spot of hunting," he wrote.

It seems their flexible lifestyle was ideally suited to exploit the arid, predator-free environment. Through the nineteenth century, they multiplied like—well, like rabbits. Fragile lands were denuded. Soil eroded away. Marsupial herbivores were driven to extinction.

The people of Australia and their government began experiments to control the rabbit population, and so began a lesson in ecology. Hunting was no match for the infestation—it required kill-

ing the rabbits one at a time. Poison was not much better, and the poisoned animals became a hazard to birds of prey that might have contributed to keeping the rabbits in check. Frank Fenner was an Australian virologist who worked in the 1930s and '40s to breed a strain of virus that would kill rabbits with a greater efficiency. His creation was the *myxoma* virus.

Myxomatosis causes pus-filled tumors to appear on the skin, progressing to blindness and death within a couple of weeks. In 1950, the Australian government approved a widespread release of the virus into the wild, with devastating effect.

For weeks, the whole countryside stank of decomposing rabbit flesh, sweet and foul and unforgettably disgusting. And everywhere on the roads and paths, rabbits staggered about, dying by inches, blind, their heads swollen and flyblown, so that it was a kindness to kill them quickly.

The rabbit population, estimated at six hundred million, dropped 90 percent in the first six months. But by then, the virus had run out of steam. First, the remaining rabbits tended to be the ones who had a higher resistance to the virus. Second, the virus in the environment led to an acquired immunity that spread through the population. And third, the virus itself mutated so that its infection was less deadly.

Another sixty years on, the rabbit population in Australia is still a national blight, and myxomatosis has become an international nuisance, against which domestic and farm animals are vaccinated all over Europe.

Aging in Predator Species

Ecologists use the word "predator" to describe herbivores as well as carnivores. All animals play the role of predator because they cannot produce their own energy and must consume other living things in order to live. So this is a fundamental difference between the ecological roles of animals and plants.

Almost all animals have limited life spans, whereas there are many plants that do not age. This is a clue that the purpose of aging has to do with ecological function. Aging in animals has evolved for the purpose of helping to protect the ecology from overexploitation. Though it is bad for the individual, aging is part of a population control program that is necessary for the community to survive even a few generations. The species that evolves to destroy the base of its ecosystem suffers a quick demise. The Demographic Theory explains hormesis and also explains why aging is nearly universal in animals though there are many plant species that don't age.

In order for the consumer species to protect the producer species, restraint in reproduction is all that is required. Aging is not a necessary piece of the solution. But the lifetime reproductive output does need to be tempered, and there is no way to do this without limiting the individual's competitiveness in the Darwinian struggle. Though the protection can be accomplished with any combination of reduced fertility and reduced life span, there's an additional advantage to arranging for relatively high fertility and dying early. Diversity and evolvability help keep the population turning over, expanding the ongoing Darwinian experiment with varying biological forms.

Tragedy of the Commons

In 1968, the ecologist Garrett Hardin published in *Science* magazine a paradigm that has since become an icon, not just in evolutionary ecology but in economics and sociology, as well. It is a rejoinder to the idea, attributed to Adam Smith in the eighteenth century, that when every individual makes wise decisions in her own self-interest, the result is an optimal world in which everyone is just as well-off as is possible—"The greatest good for the greatest number," as Jeremy Bentham would write a generation after Smith.

Hardin gave us a compelling story to help us remember that it just ain't so. Sometimes everyone can be pursuing his own rational self-interest and the collective result is a disaster for everyone.

At the edge of a small town lies a pasture where all the town residents are permitted to graze their cattle by common consent. As long as the population is small and the pasture is large, this arrangement works well. But as the town grows, the pasture begins to become crowded. Cattle are not growing as they used to because of competition for grass. Each individual farmer decides to add more of his own cattle to the pasture, compensating for the fact that his own cattle are not growing as fast or as large. But while this is a rational response for each individual, the collective result is to make the problem dramatically worse. The next year, the pasture becomes so bare that no one can find enough to eat, and all the cows die.

In his article, Hardin goes on to presage an environmental movement that had not yet been fully birthed, writing about population growth on a finite planet and the shared common resources, including the air we breathe and an ocean full of fish.

Aging in a Prey Species

The rabbit is both a prey and a predator species. Aging is useful to the rabbit as predator because it helps to protect the rabbit population from outgrowing the green plants on which it depends. And aging is also useful to the rabbit as prey. If it were not for aging, the weakest and slowest rabbits would be the small, young, defenseless ones. Imagine that the adult rabbits could maintain their strength and speed, remaining indefinitely in the prime of life. Foxes that prey on the rabbit would take immature rabbits almost exclusively, since these are the easiest to catch. The result would be that predators would allow few young ones to escape long enough to grow to maturity.

In the world where we live for real, aging is near ubiquitous, and predators routinely "prune the herd," taking the old and the sick, providing a force of natural selection that keeps the gene pool strong and growing stronger. This is much better for the long-term health of the prey species, especially compared to a situation without aging, in which population turnover is suppressed because the vast majority of the young are eaten before they have a chance to grow up. And in the long run, a large and robust prey population is also best for the predator.

Aging as a Defense Against Microbes: The Red Queen and the Black Queen

"Hermaphrodite"—the word derives from names of the Greek god Hermes and the goddess Aphrodite. Hermaphrodites have both male and female sex parts in each individual. Most flowering plants are hermaphrodites, with flowers that have both pistil (female) and stamens (male). Snails and earthworms and many other invertebrate animals are hermaphrodites, but most vertebrates have separate sexes.

Why has evolution put up with the inefficiency of separate sexes, when all creatures could be hermaphrodites and double up on their fitness?

This is a parallel question to the problem of aging, and it has been the subject of speculation among evolutionists for just as long. Sex just doesn't fit with the paradigm of neo-Darwinism; in fact, it is a bigger problem than aging, because aging typically takes about a 20 percent bite out of individual fitness, while dioecious sex (meaning with separate male and female forms) costs fully 50 percent.

The best-accepted answer to the problem of sex is called the Red Queen Hypothesis. It says that animals and plants need to change constantly in order to survive. Viruses and bacterial parasites are always on the lookout for vulnerabilities in their hosts, and since they reproduce in a matter of hours, they evolve at a prodigious rate. Bacteria in particular are optimized not just for rapid reproduction but also for gene sharing, so they are always trying out new tricks for getting a toehold in their hosts.

The Red Queen Theory says that higher organisms must continually change to defend themselves against parasites that are continually changing. Change provides a moving target to deprive the microbes of the luxury of honing their attack. Sex (as aging) also helps to keep the population diverse, and that diversity means that it is hard for epidemics to sweep through the population, because a disease that can take down one family may fail to infect the family next door.

Whether this is a correct analysis of the evolution of sex I am loath to say; it's hard to read evolution's motives from the deep past. Whatever benefit we ascribe to sex must overcome a handicap of a factor of two in individual fitness, and I daresay that without the softening of individual selection that comes from the Demographic Theory (described above), I don't think the Red Queen mechanism would stand a chance.

But what I do know is that sauce for the goose is sauce for the gander. If the Red Queen mechanism works to evolve sex, then it works even better to evolve aging.

How did sex evolve? At what point in evolutionary history did individuals learn to share their genes? I like Carl Woese's answer. Maybe life was communal from the very beginning. There were useful biomolecules and there was biological information, all shared in a pool or even in the ocean. There were no cell walls, no individuals, no competition, just different molecules struggling to make copies of themselves and each other. When cell walls evolved, when individuality and specialization appeared, competition was introduced into the recipe for life, but sharing of information and materials of life were there from the start, and of course they could not be lost, for life would not be life without them.

Evolution Evolving

The sexual sharing of genes helps to keep populations diverse and offers new combinations on which natural selection may act. Evolution in a sexual population happens faster and more efficiently than in a population that doesn't share genes. Thus sex contributes to "evolvability."

Evolvability is the most underappreciated concept in biology today. Early discussions of the subject interested only a tiny group of specialists until a well-respected evolutionary biologist teamed up with a smart and visionary computer scientist to craft a message that could not be ignored, published in the journal *Evolution* in 1996.

Darwin thought, and many after him have assumed, that the capacity to evolve is in the nature of things. It is oft stated that all you need for evolution to take place is a system that is capable of making copies of itself, with small errors that every so often provide, by chance, an advantage that enables the mutated copy to be a better reproducer than the original. College students are routinely taught that there are three necessary and sufficient conditions for natural selection to take place:

1. Variation in a trait
2. Heritability of the trait
3. The trait leads to differential reproductive success

But it turns out that there is a fourth condition, far more stringent and arcane. Gunter Wagner and Lee Altenberg argued in 1996 that the ability to evolve also depends on what is called the "genotype-phenotype map."

What is a genotype-phenotype map, and why is it so important? Here's an example. Suppose that DNA contains instructions for making an eye, and the DNA is read sequentially from one end to the other, with the body translating the instructions line by line. There are two copies of those instructions for two eyes.

A creature with two eyes close together could gain better binocular vision if those eyes were a bit farther apart. What needs to happen in order to make that evolutionary change?

If the DNA is laid out with all the instructions making an eye in-line, then the only way that the eyes could move farther apart would be to lose the present genes for one present eye and re-evolve the entire eye mechanism from scratch, one tiny mutation at a time. How inefficient evolution would be if it had to work in this way! In-line instructions may be the simplest and most economical way to code the program for constructing a body. But if the code were in-line, it would never be able to evolve. ("What, never?" "Well, hardly ever," sang Captain Corcoran.)

In fact, our DNA is not coded in-line. There is a hierarchy of genes. A complete, self-contained subroutine for making an eye is programmed into the genes. There are "hox genes" at the top of the hierarchy. These are master genes that deploy the subroutines for eyes and other body parts at the right time and place. Hox genes are like general contractors who are responsible for calling in the plumber and the carpenter and the electrician and assigning them their jobs.

The first hox gene was discovered in fruit flies, as recently as 1996. It was named "eyeless," because when this one gene was removed, flies were born with no eyes. This seemed relatively unremarkable, but what really raised eyebrows came after the technology of genetic manipulation advanced, so that extra copies of the *eyeless* gene could be spliced in. The insertion of this one gene was seen to result in an extra eye, every time it appeared. Eyes could appear on the fly's wing or leg or tail, depending only on how many *eyeless* genes were added to the genome, and where.

How did the genome come to be organized with a hierarchy of genes and genetic subroutine? In the short run, this system offers no advantage in fitness, and in fact, for the purpose of making any one organism, it is less economical and certainly less logical to have a hierarchy of genes than to have all the instructions laid out in-line.

But the hierarchical system of gene organization is capable of evolving efficiently, and the in-line system is not. Even over billions of years of evolution, the in-line system could never produce finely tuned, complex adaptations such as we see everywhere in biology.

The genotype-phenotype map is the relationship between DNA information in the gene and the body that is produced when that information is transcribed. It is also the mechanism by which the information in the DNA is "read" and translated to create a living being. Most mappings could never evolve—not in a billion years, not in a billion times a billion years. It seems that the genotype-phenotype map that we have is optimized for efficiency of evolution.

This demonstrates that evolution itself is a highly evolved process. Not only has natural selection evolved living beings that are robust, resourceful, and efficient reproducers; natural selection has also created a superbly efficient system for evolving. This is evolution of evolution, or "evolution squared," if you will.

Evolvability offers a long-term advantage to the community at a cost to the individual in the short term. This is exactly the sort of trade-off that neo-Darwinist theory says is always resolved in favor of the individual. And yet, evolvability adaptations are everywhere, built at a fundamental level into the genome, and life would not be possible without evolution of evolvability.

How Has Evolution Managed to Be So Efficient?

In a few billion years, life on Earth has evolved from small bags of chemicals to vast, integrated ecosystems of beetles and birches, plankton and porpoises. In the standard, neo-Darwinian paradigm, all this change was accomplished in tiny increments, with each increment constituting an improvement in itself. A colleague who teaches evolution likens this to a game of writing books, in which you are permitted to change or add just one letter at a time, and each version of the book must make sense on its own and must be an improvement over the version that came before it. Starting with *The Very Hungry Caterpillar*, your object is to produce *War and Peace*. The immensity of such a task, vast evolution based on tiny chemical changes (called "base pair" changes in molecular biology), reinforces the notion that evolution may have more holistic tricks up her sleeve.

Evolvability and Aging

Aging has a place in this story, because aging contributes to improving evolvability. Aging is not a make-or-break issue like the genotype-phenotype mapping. It is not even as important in this regard as sex, which changes the whole nature of the evolutionary game, making possible the evolution of tight-knit communities. But aging can make a modest quantitative difference in the pace of evolutionary change. The genotype-phenotype map has made evolution at least billions of times more efficient—maybe billions of billions. Aging is able to offer, perhaps, a factor of two, doubling the pace of evolutionary change.

So is this the reason aging has evolved? Is the purpose of aging to make evolution twice as efficient as it might otherwise be? Surprising myself, I have come around to the view that, yes, evolvability has a lot to do with the evolution of aging.

This idea is close to what is called the Weismann Theory of Aging, although Weismann never articulated it in this form and certainly never entertained the concept of "evolvability." What he did write about was the necessity of the old getting out of the way to make room in the niche for the young to grow up. The rest we must fill in for him.

Why is there a need for the young to grow up? What's wrong with just keeping the adults that we have, if the youngsters aren't fit enough to outcompete them? The answer is that constantly changing environments create the need for adaptability in the population. The population needs to turn over, to try new varieties, or it will eventually be overtaken by other populations that are growing stronger and more competitive over evolutionary time. Aging helps with population turnover. In particular, many of the young are potentially more fit than the adults they are trying to displace, but while they are small, it is hard for them to compete with a full-grown adult. Aging helps to level the playing field, and in this way, as well, it makes the evolutionary process a little more efficient. Aging also helps to keep the population diverse by placing a cap on the number of offspring that one individual can produce in a lifetime.

Everyone agrees that aging does make a significant contribution to the pace of evolutionary change. But at what cost? The usual objection to the Weismann Theory is that aging has an immediate cost to the individual—perhaps 20 percent of its fitness—and this ought to wipe out aging in short order, long before the advantage in evolvability can show any significant benefit for the future of the community. Aging exacts its cost in one individual lifetime but confers its benefit only over hundreds or thousands of generations. By then, it's just too late. Aging has long ago disappeared from the population, because those with short life spans have less opportunity to contribute to the gene pool than those with long life spans.

I believe that this conventional thinking is correct. A contribution to evolvability is not enough in itself to explain the natural selection of aging. So the Demographic Theory must come first. Nature has been forced to choose between faster reproduction and longer life span, because fast reproduction and long life spans together lead to population chaos. Once this choice is forced, then the benefit for evolvability becomes a decisive factor, leading to natural selection for shorter life spans and higher fertility, rather than the other way around.

Natural selection is not an all-out (neo-Darwinian) contest to see who can contribute more genes to the gene pool. For animals, that contest is absolutely forbidden by the requirement of fitting into an ecosystem and limiting exploitation of the producer species at the base of the food chain. The need to limit individual reproductive rates is fundamental and has been built into the evolutionary game. Under these circumstances, there is no individual "cost of aging." If lifetime reproductive output is limited by the requirement that the population cannot grow faster than the producer species, then aging is just one way to stay within this limit. There is no competition between "aging and non-aging" but only a competition between "short, fertile life plan" and "longer, less fertile life plan."

It is in that contest that evolvability contributes a decisive advantage. Aging with high fertility wins out because it evolves faster than non-aging with low fertility.

Instant Replay

Ever since Darwin, the big mystery about aging has been, what keeps life spans from growing ever longer under pressure from the intense individual competition of the Darwinian struggle? My answer to this question is that the individual Darwinian struggle has been tempered, softened by the absolute requirement that every animal species must not ride too hard the producer species on which it depends. The individual competition is for ever-faster reproduction, and this inevitably leads to a tragedy of the commons in which the foundation of the food chain is compromised, and everyone suffers. The die-off that results is swift and devastating. Beginning at the bottom, the entire ecosystem collapses and leaves bare ground where neighboring ecosystems that have evolved a better balance are able to expand and fill the void.

The result is that animals have learned long ago not to optimize their individual Darwinian fitness (which is their rate of increase) because raising the rate of reproduction beyond a critical point results in extinction, swift and sure. Aging has evolved opportunistically, in an environment where the total reproduction rate is capped by the need to protect the producer species. Thus protected from the individual cost of aging, various collective benefits are able to assure that aging evolves in almost all animals.

- For the predator, aging helps to level out the death rate in good times and in bad so that the population is less likely to overshoot its sustainable carrying capacity and come crashing down from starvation or epidemic.
- For the prey, aging helps to assure that the weakest members of the community are not the youngest but the oldest. The weakest and slowest are the ones predators grab first, and feeding the old and infirm to the predators allows the young'uns a better chance of growing successfully to adulthood.
- For any animal subject to microbial infection, aging helps to promote population turnover, keeping the population diverse so that there is better resistance to disease and epidemics cannot kill everyone at once. The faster turnover also helps the species evolving new defense mechanisms against microbes that evolve rapidly because they reproduce rapidly. If the old have weakened immune systems, they will be the first to die of an infection, and the rest of the population has a chance to develop herd immunity.

All animals and plants are evolving, and the ability to evolve and "learn" from Darwinian selection does not come free. Not every self-reproducing system is capable of evolution. Rather, the ability to evolve is itself an evolved adaptation, and aging takes its place among many, many adaptations that help populations of living things to evolve rapidly and efficiently.

NINE

Live Longer Right Now

Man lives on one quarter of what he eats. On the other three quarters live his doctors.

—FROM AN INSCRIPTION IN AN EGYPTIAN PYRAMID, C. 3800 BC

The realization that aging is self-imposed, something your body is doing to itself on purpose, yields a new perspective on health maintenance and longevity. There are things we can do to add years to our lives and things we can do to be healthier in the present, and fortunately for us, these are mostly the same things. A program for life extension* is likely to make you feel better in the present and even help you get sick less frequently.

Much of what I have to recommend for self-care is already standard medical advice. Exercise, weight loss, and a daily aspirin or ibuprofen are among the best things you can do for yourself, and I'm sure you didn't hear that first from me. But there is also something new in our program to cheat the Black Queen. The most difficult conceptual leap I ask of you is to question all reverence for the natural. I grew up with the counterculture and celebrated the first Earth Day when I was a college student. Culturally and socially, I feel at home with the crunchy granola crowd, so imagine a lilt of sadness in my voice as I tell you that "natural" has little to offer for life extension.

All that I recommend is based on studies of people or rodents. I don't believe that pure theory is a good guide in medicine, because there is

* My Web site at AgingAdvice.org summarizes a program for health and longevity.

just too much that we don't understand. And I don't have much faith in supplements or drugs that work for lab worms or fruit flies because, as it turns out, it is just too easy to increase life span in these simpler animals, using tricks that won't work for you and me. But what works for lab mice is usually, not always, going to offer benefits for us, as well.

Substances that have been found to lower mortality rates in humans are anti-inflammatories (such as aspirin and ibuprofen), vitamin D, and the diabetes drug metformin (Glucophage). Fish oil and turmeric are natural anti-inflammatories that have been associated with protection from heart disease, stroke, and dementia. Substances that increase life span when fed to rodents include metformin, melatonin, and deprenyl (Selegiline). Rapamycin is the most recent and most powerful of the drugs that extend life span in mice, but it is likely to leave us vulnerable to a lot of infectious diseases, and I don't recommend it.

Vitamin D is in a class by itself. High blood levels of vitamin D are associated with lower risk of cancer and infectious diseases, and no one really understands why.

Telomerase activation—the turning on of the genes to resume production of this biologically rationed enzyme needed for sustained cellular reproduction—is a promising idea for the future, but what is available now is not very effective. Still, it might be worth adding to your regimen.

A low-carb diet coupled with periods of intermittent fasting provides the easiest way to fool the body into thinking it is getting less nourishment than you are actually eating, with likely benefits for health and longevity.

People who are happy, passionate about their work, and engaged daily with friends and family live a lot longer than people who are depressed and isolated. Share your gifts with others and you will have a long and satisfying life. This is no small thing.

A Common Misconception About Life Extension

Many people hear about the life extension program and imagine forestalling the end after the body is already depleted and there is not much life left in us. But antiaging science is not about tacking on

extra years in the nursing home. It's a program for retaining the vitality, adaptability, and youthful capacity to learn and to grow during years when we might be expecting to be losing all these faculties.

Deconstructing the "Natural"

Most of us can't remember a time before the meta-marketing phenomenon of "natural." But fifty years ago, technology was king, and we had no compunctions about improving on nature. In the 1950s, tonsils were ripped from the throats of small children because they had a tendency to turn red during laryngeal infections, so doctors thought nature had made a mistake. In the 1950s, Dr. Spock had to break with standard medical advice to recommend breast-feeding over infant formula. And don't forget that Wonder Bread helped build strong bodies twelve ways. For half a century, we have been told about natural foods, cosmetics, soaps, herbal remedies, and even items of clothing. Natural=healthy. The medical establishment—much to its credit—has learned to respect the body and work with it to promote natural healing, rather than rush to fix what ain't broke. Today, natural treatments for every disease are often presumed to be preferable whenever such are available.

So far so good, but it takes some reflection for us to take the next step. We must acclimate to a different reality about aging: natural diets, herbs, and remedies are unlikely to slow the aging process.

This book has argued that aging is not a bug in evolution's program but a design feature that is naturally selected in its own right. Aging is "natural" in the deepest sense, that it is a product of evolution, built into our genes. At root, the appeal of the natural comes from faith in evolution—what is natural is part of the environment in which humans and our ancestors evolved; hence we are presumed to be well adapted to it. If natural foods are better for us, it is because they are the foods that evolution has equipped our bodies to work with. (Follow this logic a step further and you reach the "paleo diets" that try to mirror ancestral food choices.) Natural selection has not prepared us for the pace of life in the jet age or for breathing smog or drinking Coca-Cola; hence many of the complaints of modern life may be attributed to a mismatch between the life we are living and the life for which evolution has prepared us.

And indeed, it is likely true that many of our ailments are products of modernity: lung cancer from cigarettes and urban smog, metabolic syndrome (increased fat, blood pressure, blood sugar, and other factors leading to type 2 diabetes) from junk food, nervous disorders from overstimulation, and depression from living in a fragmented and disconnected society.

Getting and spending, we lay waste our powers:
Little we see in Nature that is ours

William Wordsworth penned these lines in 1802! What would he say about our mode of life in the twenty-first century? The damage that we suffer because of the isolation created by modern Western cultures, the chemical pollutants and the packaged diet, the imposed schedules that override our psychological needs and our biorhythms—all these are very real and harmful. But they have little to do with aging.

Aging is not a disease of modern creation. It was not born yesterday; it has a hoary legacy. Just take a look at photographs of people from the nineteenth century, or consider the ages of people portrayed as old in Victorian novels. These are people that lived before cigarettes and pesticides and junk food and the fragmentation of communities. By today's standards, you could say that everyone in the nineteenth century was eating a natural diet, and yet they looked and felt and acted older than people of the same age today. In nineteenth-century novels, the forty-year-olds have lost their bounce; fifty-year-olds look like today's seniors, and of course, few people lived past sixty. Life expectancy was much shorter in the nineteenth century, and it was not only because mothers died in childbirth and infectious diseases killed people in their prime. People used to lose their health and vitality and cognitive function at ages that we now think of as the prime of life.

The idea that natural is good for us tacitly devolves from the notion that evolution has put us together in such a way as to optimize our health. By providing a natural diet or natural artifacts, we imagine we are helping the body to function as it was designed to by natural selection. We get out of the body's way and allow it to do its best healing for us. This presumption holds well for the ailments and complaints of youth. But if you have come with me so far through this book, I expect you may just be wondering whether aging is in fact a genetic program of self-destruction. In this case, the body is not doing its best to heal—quite the contrary, the body is working against itself. Natural foods and remedies can only help the body better to destroy itself.

Hormesis, and Tricks to Keep the Pounds Off

Caloric restriction is the oldest, most reliable, and universally recognized way to extend life span in lab animals. Growing numbers of people are trying to apply this approach for their own health, vitality, and longevity. But many people find the deprivation impossible to sustain in the long run, and even those who keep the discipline find that hunger can make them edgy and irascible. The joke on the street is, "It may not make you live longer, but it sure *seems* longer."

So there is a market for drugs and supplements that are able to simulate the hormonal activity of a body deprived of calories, even while fully fed.

The main way that the body knows whether it is over- or underfed is through the hormone insulin. Whole textbooks have been written on the subject, but the CliffsNotes version is this: when the body is exercising a lot and eating a little, fat is broken down to make sugar, and sugar is pumped into the blood to fuel the muscles. This is the low-insulin condition. But when the body is eating a lot and exercising a little, sugar accumulates in the blood, so insulin is circulated in order to signal the body to turn sugar into fat, which is stored for a rainy day. This same insulin is also telling the body that there is plenty of food around, so it's a good time to reproduce and die. Low insulin is the body's signal that food is scarce; it is a bad time to reproduce, but the right time to slow down aging so that the body will be alive and youthful, ready for reproduction when the famine is over.

Starches and sweets turn quickly to sugar in the blood, causing a surge of insulin. Every bowl of pasta sends a message to the body to put on body fat and accelerate the aging process. It may be healthier to avoid carbohydrates, and this may be a way to get some of the benefits of calorie restriction without actually eating fewer calories. Most people find that after a low-carb meal, they remain satisfied longer, and hunger for the next meal comes later. These benefits are the basis of low-carb diets, popularized by Robert Atkins and Barry Sears, and a generation earlier by Herman Taller.

Strange to say, artificial sweeteners seem to trick the body in the other direction. The insulin metabolism is fooled by sucralose and saccharine, and some studies report that their effect is worse than sugar itself.

In middle age, the insulin metabolism slowly turns against us. As more insulin is circulated, the body loses sensitivity to insulin, so ever more is required to keep blood sugar in the optimal range. This is insulin resistance, or type 2 diabetes, or metabolic syndrome, and it is a hallmark of aging in humans. Everyone is affected. Gradual loss of insulin sensitivity and prediabetic symptoms in middle age are considered "normal."

The standard treatment for loss of insulin sensitivity, prescribed to diabetics for over fifty years now, is metformin. It is long out of patent and inexpensive. Because it has been taken by tens of millions of patients, there is a great deal of experience and epidemiological data on metformin. Thus it was discovered by accident that people taking metformin for diabetes have lower risk of heart disease and some cancers. This has inspired speculation that metformin is an antiaging drug. Diabetics who take metformin live much longer than untreated diabetics or diabetics on more "modern" drugs. Two recent studies (from Scotland and Wales) found that diabetics taking metformin had lower mortality rates than nondiabetics who don't take metformin! Ordinary mice, not overweight or diabetic, live up to 35 percent longer with metformin. It may turn out that almost everyone can benefit from metformin starting around age fifty.

Aging to Be Addressed as a Disease

In the spring of 2015, Nir Barzilai of Einstein Medical College in New York received FDA approval for the first clinical trial of a

drug targeted not at any particular disease of old age but at aging itself. The drug being tested was metformin.

How will he know if it is working? Barzilai's protocol is to administer the drug to people with Alzheimer's symptoms and then see if it lowers their risk factors for heart disease: conversely, he will offer metformin to heart patients and see if it slows the age-related loss of cognitive function. The plan also calls for monitoring of signal molecules in the blood that have been associated with age.

For those who prefer to stay away from prescription drugs, there are naturally derived supplements that also can improve insulin sensitivity. Berberine (from goldenseal) is an herbal extract with a 5,000-year pedigree in Oriental medicine, which produces a better metabolic response than metformin in short-term human trials. Magnesium is a chemical element, a mineral most of us don't get enough of. Pycnogenol is extracted from the bark of a French pine tree. Resveratrol is derived from wine grapes. Chromium, a trace mineral, may also be useful in blunting the insulin spike. *Gynostemma pentaphyllum* is the traditional Chinese herb *jiogulan*, which has been promoted recently for its ability to enhance insulin sensitivity.

There are things you can do just before meals that tell the body that this food is to be burned, not stored as fat. There is evidence that garcinia and irvingia, taken twenty minutes before a meal, work in this way. Cinnamon and vinegar might also be effective. (Green coffee extract has been promoted for this purpose, but a prominent study proving its effectiveness was withdrawn because of fake data.) Simply drinking a glass of water before every meal is a way to slow down your food absorption and add to the feeling of fullness. My personal favorite is a minute or two of high-intensity exercise right before eating. You can do push-ups or pull-ups or jumping jacks or climb stairs or run around the block or skip rope "red hot pepper" style.

Caloric restriction mimesis is a kind of low-hanging fruit in antiaging medicine. It is likely to provide real results in the near term, but its benefits are limited. Estimates range from three to ten years of life extension by this pathway. (The 120-Year Diet of Roy Walford is now understood to be a major exaggeration.) If you lose weight and take metformin and resveratrol and fast once a week, this is like adding the same three years over and over; the benefit is not likely to be additive.

Hidden Exercise

Going to space is an iconic dream—a combination of leaving humanity's primeval cradle but also, in a more physical sense, returning to a womb where floating, rather than walking, is the norm. Astronauts in space have been dumbfounded by the blue beauty that is Earth, seeing it rise from the moon or, in lower orbit, marveling at circling the planet every ninety minutes, seeing the sun cut through the thin ribbon of the atmosphere, sending their white cabin briefly through all the colors of the rainbow—and then into the Big Night, with Earth visible only as the place where there are no stars.

But the flip side of such soul growth in space is the peril to bodies not exposed to the normal, healthful stresses of gravity. NASA scientists have estimated that a ten-month mission to Mars would expose thirty- to fifty-year-old cosmonauts to such muscle deterioration due to lack of gravity that they would arrive as weak as eighty-year-olds, too weak to walk about the Red Planet in spacesuits. On the International Space Station, crewmembers lost on average 15 percent of their muscle mass and 25 percent of their strength, even though they were exercising regularly.

Often the greatest windfall of a trip is the new light it sheds on home when we return. Space missions drive home a physiological lesson for our new understanding of aging: our youthful physiques are maintained not just by nutrition and by grace of not living too long but crucially by the unseen, unfelt, normal stressors of our Big Blue Exercise Machine.

It's Not About Calories

"The amount you weigh is exactly the difference between the calories you ingest and the calories that you burn exercising." We hear this all too often. Writers who should know better promote it as the First Law of Thermodynamics. This is dangerous nonsense.

In fact, the calorie content of a food is measured simply by burning it and collecting the heat that is released. But the body's efficiency in use of foods is a very complex affair, dependent on everything from how well you chew your food to which bacteria reside in your intestine. In the charts, peanuts have exactly the same caloric content, pound for pound, as peanut butter, but in reality you absorb a lot more calories from the peanut butter. There are (lucky?) people with very inefficient metabolisms and (unlucky?) people whose bodies are able to extract every last calorie from any meal. The bacteria that live in our guts digest food for us but exact a toll in energy that they need for themselves. Depending on the particular bacteria you have in your intestine, the toll may be only 10 percent of the food energy coming in, or close to half.

Roughage slows calorie absorption and helps to move food quickly through the intestine, with less total absorption. Eat as much raw wheat bran as you can stomach, and think of it as a negative-calorie food. A vegetarian raw-foods diet is not for everyone, but if you can live with it, it is a sure way to lose weight. Raw foods are poorly absorbed, and from a Black Queen perspective, that's a good thing.

In experiments with flies and worms on a low-calorie diet, just the *smell* of food undid some of the benefits of caloric restriction. This effect has not yet been replicated in mice or rats, but in experiments with humans, insulin levels in the blood can rise when people smell or see food or even just think about it. I know people who chronically struggle with their weight who quip that "just looking at food makes me fat." Strange as it seems, there may be some truth in that! Lab worms exposed to the scent of their food, yeast paste, experience reductions in longevity; if the aging mechanism in them is phylogenetically conserved, it may be similar in us. If these animal experiments can be extrapolated to humans, it may be best to avoid obsessive thinking about food, and stay away from delicious smells. There may be risks to working in a bakery even for those who can resist the temptation to sample.

Fasting

With any diet, the most important thing is to find a regimen that you can live with. People who try too hard to restrict calories end up gaining weight, and a series of sobering studies from California tells us that's about 90 percent of people who diet. It's good to have a variety of options available so you can try them and see what habits you can sustain.

Some people find that they have no trouble with discipline as long as they know it is temporary. Good results have been obtained, both with humans and with mice, fasting every other day. Both people and mice tend to make up the difference by eating two days' worth on the alternate days. So fasting turns out to be largely ineffective for weight loss. But for health and a youthful insulin metabolism, results from intermittent fasting are almost as good as if we were actually eating less. And mice that eat every other day live almost as long as mice that are calorically restricted.

I fast one day a week. Nothing but water and maybe some sugar in my tea from Wednesday bedtime to Friday breakfast. I also try not to eat for a few hours before bedtime and a few hours after waking in the morning. I've grown accustomed to this schedule over seventeen years now, and I no longer find it difficult.

Valter Longo of the University of Southern California has studied the effect of the four-day fast in people and in mice, and he reports that it has a remarkable rejuvenating effect on the immune system, clearing the white blood cells that haven't been useful in many years and adding a new crop of naïve T cells, primed to go after new exposures, which is exactly what we tend to lose with age.

You may remember Longo from chapter 6. As a grad student, he discovered cell suicide in yeast, and he persisted for a decade before the scientific community would finally acknowledge that what he reported was real. In recent years, Longo's biggest project has been to document the benefits of fasting for cancer patients. A three-day fast prior to chemotherapy almost entirely relieves the nausea, fatigue, and headaches associated with chemo while making the therapy many times more effective against the cancer. Being starved of food turns on protective mechanisms in normal cells, so they are less vulnerable to the assault of chemotherapy; but cancer cells behave in exactly the opposite way and are primed by starvation for the kill. Recent studies also show that a ketogenic diet, high in fat with almost no carbs, can destroy cancer tumors in mice.

Personally, I was frightened by the idea of a four-day fast. That's why I tried it. It turned out to be not as difficult as I feared. My mind is slowed down when I am not eating, and I notice I have trouble putting ideas into words. I can't run or swim on fasting days, but I can still cycle and hike and do yoga. For me, fast days are frequently a time of creativity and wide-ranging ideas.

It's counterintuitive but true that health and longevity are better served by clumping our food consumption (feast and famine) than by spreading food consumption evenly through the day and through the week. This topic is still controversial, and there is unclear, even contradictory evidence being reported. My bottom line recommendation is that it is worth trying. Experiment with different schedules, because individual response varies widely.

Can You Eat While Fasting?

Fasting on water isn't for everyone. It requires willpower for most of us, and for many it entails a lull in motivation and productivity. Can we reap the benefits of fasting without the hunger and the disruption to our lives?

Based on interviews and physiological measurements of human subjects, Longo developed a 5-day program that he called the Fasting-Mimicking Diet (FMD). In the spring of 2015, he published

his first results, and announced a company called L-Nutra that would sell packaged, prepared vegan foods from natural sources, pre-measured for a 5-day FMD.

The diet is low in calories and very low in protein. It is probably not healthy to maintain it for more than a few days at a time, but Longo claims that the 5-day program produced much the same physiologic response as a 4-day water fast. The formula is based on meals of about 360 calories, derived 9% from protein, 44% from fat, and 47% from carbohydrates. There are 3 such meals on the first day (1090 calories) and 2 on each of the next 4 days (725 calories).

Enid Kassner and I have developed a series of 360-calorie vegan recipes on the Longo proportions, available at FMDrecipes.org.

Exercise!

. . . and more exercise, activity, and working out, and just keep moving. It's all good—long-distance jogging and short sprinting, swimming, yoga stretches, strength training, and especially intense bursts of effort, styled as "interval training."

Build exercise into your schedule. You may find you can bicycle to work in less time than it takes you to drive. Climbing stairs instead of taking the elevator can be a convenient mode of interval training. Abandon the coffee shop and meet your friend for a walk in the park instead. If you sit in an office chair all day, consider investing in a treadmill desk, or simply transfer your notebook to a shelf so you can alternate fifteen minutes of sitting with fifteen minutes of standing. Be creative, go for it—the more the better! Health benefits of exercise are broad and deep. Exercise is the best mood elevator that money can't buy, the only antidepressant that doesn't lose its potency over years of application. Exercise releases endorphins, hormones that work as natural painkillers. Exercise improves your resistance to disease in the near term and extends your life expectancy in the long run.

Gym rats live longer than couch potatoes, and elite athletes live longer than gym rats. In lab experiments with exercise wheels, mice that run all day even while half-starved live the longest. My advice is to exercise as much as you can tolerate, and then up the standard as your endurance improves. For those who are making the break from sedentary habits, remember that a little exercise is a whole lot better than none.

Yoga is exercise with attention on body sensations. You may think of it as a way to retain balance and flexibility as you age, and it is all that and more. Consistent attention leads to learning leads to mastery. Out of India comes an ancient tradition, masters who learn to control their breathing, and then their heartbeats, and then their body temperature, aspects of metabolism that Western science regards as outside of conscious control. There are credible tales of yogis who learn sufficient control as to maintain health and slow the aging process. If this path appeals to you, it is a powerful one. Some of the same benefits for metabolic control might be realized directly through biofeedback training.

Many older people find that arthritic pain limits their physical activity. The paradox is that exercise is one of the best things you can do for arthritis. Find ways to move that you can tolerate, and then expand the limits as you become healthier. Back pain responds swimmingly to laps in the pool.

For many years (based only on common sense), endurance exercise was thought to be the most beneficial for health and life exertion. But in the 2000s, this wisdom has been turned on its head, and the new thinking is based on evidence rather than intuition. Short bursts of high-intensity exercise are more effective for your health. Sprint until you feel that you just can't get enough air. A few repetitions with a heavy weight are more effective than many repetitions with less weight. Choose an exercise that you can repeat five to ten times before your muscles fail.

A four-minute workout at maximum heart rate stretches your cardiovascular limits better than an hour of jogging. You can take any four-minute exercise and accelerate the pace by 20 percent, and then see if you can tolerate it for ninety seconds. A good sprint should leave you still panting and gasping for breath two minutes after it's done. Better than a cup of coffee, it can be an afternoon pick-me-up for those with a little (okay—maybe a lot of) willpower. For those who can tolerate it, calisthenics and weight training at the limit of your strength hold powerful benefits.

The caveat is that this judgment is based on physiological markers and not on life expectancy. It takes too long to see which regimens do the most to improve long-term survival, so researchers use lab measurements as a surrogate. Intriguingly, high-intensity exercise also releases bursts of beneficial hormones. Blood pressure, insulin sensitivity, muscle mass, fat loss, lung capacity, heart rate—by all these criteria, high-intensity exercise seems to be best.

An Exercise Program to Fit Your Temperament

What if statistics say one thing and your body says another? Listen to your body! Forcing yourself to adopt an exercise program that you hate is unlikely to take you very far. High-intensity exercise is great if you can tolerate it, but please don't beat yourself up if it is not the right style for you. Take the dog for a walk around the block and pat yourself on the back for a job well done.

In the long run, exercise depends on habits rather than willpower. Anyone can benefit from more physical activity, and the way to get there is not to make New Year's resolutions but to gradually build new routines over time. Exercise benefits everything from skin complexion to insomnia, helps prevent twelve kinds of cancer, and is the number one line of defense against stroke, heart disease, and diabetes.

If you come to enjoy exercise, you will sustain the practice, and the practice will sustain you.

Why does exercise work so well? Perhaps we don't think to ask this question, because we've been familiar with the benefits for so long that they no longer seem strange to us. But in the context of any of the wear-and-tear theories of aging, the benefits of exercise are a paradox. Not only is there (short-term) physical damage from every form of exercise, but exercise dramatically increases oxidative metabolism and promotes damage from free radicals. From the perspective of the popular Disposable Soma Theory, exercise burns calories and makes less energy available for repair. Theory predicts it should shorten life span. The only theoretical idea that can account for the broad benefits of exercise is hormesis. Exercise is second only to starvation as the healthiest poison known to man.

There's one curious caveat. In lab tests with groups of animals, the group that exercises consistently has a greater average life span than the control group, but there are one or two animals in the control group that beat the odds; without exercise, they live just as long as the longest living animals that do exercise. This suggests that benefit from exercise is subject to genetic variation and that there are some individuals who don't need to exercise in order to achieve full life extension. Ask a centenarian how she came to beat the odds, and the chances are she won't say a word about exercise. Less than 2 percent of us fall in that lucky group, and, the way our minds play tricks, 70 percent of us like to imagine we are among those 2 percent!

Hormesis is about stressing the body, so that the body overcompensates and becomes stronger. This perspective goes far to explain why exercise is broadly beneficial even though it is at best an acquired taste, and for most of us, it feels uncomfortable every time we do it. Really vigorous, gut-wrenching exertion inspires more avoidance behaviors than just about anything we can attempt. Most men who successfully overcome the aversion invoke the competitive instincts and exercise in the context of contests with others; most women who achieve consistency in their exercise program do it by taking classes and forming support groups. Personally, my style runs with the feminine, but I'll cop to a surge of adrenaline when a younger man passes me on the bike path.

Four Ways the Body Destroys Itself

As we age, our bodies are not simply failing to maintain themselves; they also engage in active self-destruction. There are four main mechanisms by which this occurs:

- Inflammation, which in youth is narrowly targeted at external invaders, loses its discrimination and turns on the self in old age. Inflammation attacks joints (arthritis), arteries (atherosclerosis), and neurons (Alzheimer's), and is statistically associated with many types of cancer.
- The thymus gland, where T cells are trained, atrophies steadily with age. The immune system becomes less effective against invading viruses and bacteria. *Complement* is a "first responder" component of the immune system that has been linked to arthritis and to macular degeneration of the retina in later life.
- Apoptosis effectively removes damaged cells in youth, but in old age, some rogue cells are missed, leading to cancer. This is a "Type I" error, in which apoptosis has failed to do its job. Equally important are "Type II" failures, which result in loss of healthy, functional cells to apoptosis, crucially including nerve and muscle cells.
- Telomere shortening—cellular senescence is a double-barreled attack on the body. Tissue repair and renewal are slowed as stem cells become fewer and less active. And cells with short telomeres are a source of pro-inflammatory signals that derange other aspects of the body's operation.

Aspirin, Fish, and Curry—An Anti-Inflammatory Program

Inflammation is an important first-line defense against infection, but as the body ages, inflammation is deployed indiscriminately, helping to trigger all the diseases of age. The reason to suspect that inflammation is an affirmative "suicide adaptation" is that even dumb anti-inflammatory agents are able to substantially extend life span, though they do nothing to distinguish legitimate targets of inflammation from self-destruction. Aspirin, ibuprofen, and naproxen (NSAIDs—the acronym is for nonsteroidal anti-inflammatory drugs) simply dial down the body's inflammatory response in general, literally protecting us from ourselves; nevertheless, NSAIDs taken daily are associated with lower risk of heart attack, stroke, dementia, several types of cancer, and probably Parkinson's disease, as well. It is estimated that daily doses of aspirin can add three years to your life span—the only proven life extension in humans to date from a drug or supplement!

A small percentage of people have digestive systems that can't tolerate aspirin or ibuprofen or both. If you are one of them, you will know in short order. Experiment with different NSAIDs, and use doses lower than those recommended for, say, headache relief. There is no evidence that more is better. Don't mix aspirin with alcohol or vitamin C. Don't take both aspirin and ibuprofen on the same day. Talk to your doctor. Be prepared to give up on NSAIDs if you can't find one you can tolerate.

In recent decades, inflammation has been discovered as a prime culprit in all the diseases of old age. Not long ago, the sense of the medical community was that cardiovascular diseases were caused by a gradual buildup of cholesterol deposits obstructing the arteries. In the 1990s, the perspective began to emerge that now dominates the field: atherosclerosis is primarily an inflammatory disease. Heart attacks and strokes are triggered when hot spots of inflammation in the arterial wall break loose and lodge elsewhere in the circulatory system. Even more than cholesterol-lowering drugs, anti-inflammatory agents are our best pharmaceutical defense. Ironically, the original motivation for prescribing daily aspirin to heart patients was to "thin the blood" and slow clotting that can lead to obstruction. It was luck and not understanding that brought low-dose aspirin into widespread use long before its primary benefit (as an anti-inflammatory agent) was appreciated.

Likewise, it was not suspected that daily NSAIDs would lower the risk of cancer until this fact emerged from statistical studies. People who were taking NSAIDs to hold down risk of heart attack surprised epidemiologists with lower cancer rates, as well. Anti-inflammatory agents have been shown to protect against breast cancer, colorectal and stomach cancers, esophageal cancer, lymphoma, and other cancers.

Similarly, no theoretical link was suspected between inflammation and dementia until it was discovered after the fact that heart patients taking NSAIDs received significant incidental protection from Alzheimer's disease.

Many people find that daily aspirin or ibuprofen reduces the pain and stiffness of arthritis. This should be no surprise. Up until a few years ago, the cause of osteoarthritis was described as cartilage wearing away during a lifetime of abrasion. Osteoarthritis was contrasted to rheumatoid arthritis, which commonly occurs at younger ages, and was recognized as an autoimmune disease. But the distinction has become blurred, as increasingly it is recognized that osteoarthritis, too, is a result of the body attacking its own cartilage and spinal disks with inflammation.

Natural anti-inflammatory supplements include fish oil and turmeric (curcumin). Curcumin constitutes less than 5 percent of whole turmeric, and it is poorly absorbed into the blood. There are several formulations on the market that concentrate curcumin and claim to make it more bioavailable. But there is also epidemiological evidence that people who just eat a lot of curry have lower risk for some of the diseases of old age.

It is speculated that any benefit of prescription statins has more to do with their anti-inflammatory action than their effect on cholesterol. You may find ways to lower cardiovascular risk that have fewer side effects than statins.

Melatonin and the Body's Clock

Melatonin is a hormone closely connected to the body's daily clock, and we may speculate that it is related to the long-term aging clock, as well. It is a popular drugstore item, as travelers find it useful for resetting the body clock when they cross time zones. Melatonin can help tell the body it's time for sleeping, but the effects are far broader.

Melatonin is produced in the brain (specifically the pineal gland) and then circulates through the blood to affect the whole body. It is a high-level signaling hormone that affects gene transcription, with a cascade of lower-level effects.

There is good evidence that supplementing with melatonin works to modestly extend rodents' life spans. Melatonin is an antioxidant, but in the tiny quantities that the body deploys it, this is probably not significant. Young people generate melatonin in their bloodstream around the same time each evening, signaling the body to prepare for sleep; and melatonin levels stay high through the night, dropping off before it's time to wake up. Older people have less melatonin at night. Greater numbers of older people than younger people suffer from sleep disorders, and many older people take melatonin at bedtime just because it helps them sleep.

In my personal experience, I find that 1 mg of melatonin helps me fall asleep at night. Side effects include morning "sand" in my eyes, exacerbation of apnea, and possibly an effect on dreaming—it's hard to tell. I find that if I take more than 1 mg, the side effects are worsened, and I have trouble waking in the morning.

Although all the data is not in, it does appear that melatonin, taken as a supplement, slows the spread of cancer, strengthens the immune system, and retards the aging process. Vladimir Anisimov, a Russian biochemist who has studied a large number of antiaging interventions in his lab, is an eternal optimist on antiaging hormones and supplements. Many of the studies associating a longer life span with melatonin come from Anisimov's research group.

Melatonin is a potent neuroprotective agent, demonstrated in tests where animals' brains are subjected to ischemia (oxygen deprivation). It has been proposed as helpful for Alzheimer's treatment, and for Parkinson's there is preliminary data, though clinical evidence remains uncertain. Taken orally, it is easily absorbed and quickly boosts blood levels. Melatonin is cheap and convenient. In the mid-1990s, several books were published promoting broad benefits. Walter Pierpaoli led the charge. There followed the inevitable backlash, but when I review the warnings in these papers now, I find they have little substance. The worst they had to say was that melatonin needs more study. This remains true today, and melatonin's chief drawback in this area is that it is unpatentable and too cheap to motivate any business to invest in research.

Preserving the Immune System

The immune system becomes less effective with age, and old people commonly die of infections (like pneumonia and influenza) that would not be dangerous in a younger person. The immune system also plays an important role as a sentinel against cancer cells, destroying them one by one before they can grow into visible tumors. Thus weakening of immune function also contributes to the rising burden of cancer with age.

These are errors of Type I—failure of the immune system to recognize and attack the body's enemies. But even more insidious are the Type II errors, the "false positives" in which the immune system turns upon the body's own tissues and attacks them from within. Type II errors are behind the inflammation that is linked to every disease of old age. In addition, a growing percentage of people over age sixty have antibodies against the self circulating in their blood, and about 5 percent have symptoms identified conventionally as belonging to autoimmune diseases, such as lupus, inflammatory bowel disease, arthritis, asthma, Hashimoto's thyroiditis, and MS. Herbs that support immune function include reishi mushroom (Chinese *Lingh Zhi*, genus *Ganoderma*) and black cumin seed, a.k.a. charnushka or kalonji or *Nigella sativa*. Some people find reishi mushroom effective for short-term boosts to the immune system, when they think they're coming down with something. Astragalus root and organic aloe juice are also sometimes considered general immune system stimulants. So are beta-glucans, found in mushrooms and oatmeal but available also in pill form.

But the most effective and best documented supplement to support immune function is Vitamin D, and most of us can benefit from jumbo doses.

Vitamin D

We were all black when we came out of Africa several hundred thousand years ago, but skin lightened for those that moved north. The reason for whiter skin was likely to allow more sunlight in, to obtain more vitamin D.

Supplementing with huge doses of vitamin D might be the best thing you can do to preserve immune function with age. High blood levels of vitamin D are associated with lower risk of infectious disease and most types of cancer. Autoimmune diseases from arthritis to asthma to multiple sclerosis to lupus sometimes respond to high doses of vitamin D (mostly anecdotal evidence, some statistics). Vitamin D is known to aid in preventing bone loss, and there is even evidence that vitamin D can slow the progression of age-related insulin resistance, or type 2 diabetes.

Sunlight is the best source of vitamin D, but the sun causes skin to age and can lead to cancer. It's hard to get enough D from sunlight in the winter. If you rely on sun for vitamin D in the summer, try to expose your whole body for shorter periods, rather than just your face and arms for longer periods. UV tanning lights can also have a place in a vitamin D program, if used in moderation and applied to areas of the body that don't get much sun. (Watch out to avoid burns, however, which correlate more directly to skin cancer than just being exposed to sunlight.)

There seems to be a disparity between the recommended daily allowance and the recommended blood level. The RDA of vitamin D is still only 600 IU (up from 400, where it stood for many years). A dose of 600 IU is a tiny dose, equal to 15 μg. That's fifteen-millionths of a gram. The recommended blood level is 20–100 ng/ml, and there is good reason to aim for the high end of that range. Many people find that they can only get their blood level up into the 80s or 90s by taking 10,000 IU or more daily. My recommendation is to have your blood level checked the next time you have a physical, and if it is low, don't hesitate to take 20,000 or 30,000 IU daily until your blood level comes up. That's fifty times the RDA, but it's still tiny compared to your daily dosage of vitamin B or C. There is little cause for alarm if your blood level rises above 100, and for some people, it might be beneficial.

The only potential downside of high-dose vitamin D is that it tends to raise levels of calcium in the blood. Adding vitamin K to the mix can prevent this. The forms you should look for in supplement pills are D3 and K2 (which are usually what you'll find).

Extending Telomeres

There is a real possibility that telomere length constitutes a primary aging clock of the body and that lengthening telomeres will restore youthful function and add years to life expectancy. Telomeres are the most promising target for antiaging intervention in today's research. However, the race to find effective treatments is cloaked in corporate secrecy and ridiculously underfunded. At the same time, it is hamstrung by patent law and FDA regulation. The situation has become an absurd and infuriating collision of science with capitalism.

We first discussed telomeres in chapter 5. A telomere is a cap of repetitive DNA attached to every chromosome, and each time a cell divides, its telomeres get a little shorter, keeping count of replications. When the telomeres become too short, the cell slows down and stops reproducing. Worse yet, the cell can become toxic to the rest of the body, secreting signals that increase inflammation and accelerate apoptotic self-destruction.

There is a plausible hypothesis that telomere length in stem cells is the basis for one of the body's primary aging clocks. Short telomeres may be a principal reason why old bodies behave differently from young bodies. Enough people believe in this idea that there are hundreds of thousands of people taking herbal supplements that have only a marginal effect in lengthening their telomeres. And many biotech companies are in a race to develop drugs or herbs that do a better job.

The machinery to rebuild telomeres is already in the body, in the nucleus of every cell, in the form of genes coding for the telomerase enzyme. Telomerase is the right tool for the right job. But we can't simply eat telomerase or even inject it into our veins. The large and delicate molecule does not survive digestion and will not pass from the blood into the cell nuclei where it is needed. (If it did, many more people might be ordering *ikura, tobiko,* and *masago*—salmon, flying fish, and capelin roe, respectively—fish eggs with plenty of telomerase.) In nature, telomerase is created within the cell for its own private use, not circulated through the body. But in adults, the telomerase gene is turned off almost all the time, so our cells produce very little telomerase after we leave the womb. The consensus strategy for lengthening telomeres is to activate the cell's own telomerase genes, using a molecule that can pass easily through the body into the stem cell, through the cell, and into the nucleus, where it finds the gene for telomerase and turns it on. This is a "gene promoter," also called a transcription factor.

There are a handful of herbal products now on sale that claim to be able to turn on the telomerase gene. A short peptide (protein) called carnosine has some of this effect, and herbs that are reported to turn on the telomerase gene include astragalus, ashwagandha, bacopa, boswellia, green tea, horny goat weed, and milk thistle. The frustrating thing is that companies that sell these supplements don't have enough funding to study the effectiveness of their products in detail. And of course Big Pharma is not interested, because herbal extracts cannot be patented, so there would be no way for corporate giants to capture a return on their giant investments.

The two companies that have actually published research studies on the subject are Geron Corp and T.A. Sciences. Geron is the oldest and best-established biotech company devoted to aging science, and they performed some foundational research in the 1990s but then abandoned telomere research because they didn't think they could make enough profit. T.A. Sciences is a small New York company that bought the rights to the best-performing extract that came out of Geron's research. They have published two peer-reviewed articles by prominent researchers documenting the benefits of their product, TA-65. Though T.A. Sciences will not tell us what ingredients are in TA-65, it has been analyzed from purchased samples of the pills, and the active substance is widely reported to be cycloastragenol, a trace ingredient that must be concentrated from much larger quantities of astragalus root. Luckily, astragalus is cheap and easy to grow, but cycloastragenol, so far, is expensive and difficult to extract.

We don't know that cycloastragenol is the best of the herbal ingredients that promote telomerase. Bill Andrews is a credible scientist in the field who did the pioneering work on telomerase at Geron. His company, Sierra Sciences in Reno, has a very efficient cell test that has been used to screen hundreds of thousands of chemicals to see which might be effective at turning on telomerase. Andrews has told me that astragalus extracts are far from the best herbal products now available for turning on telomerase and that TA-65 works too well to be merely cycloastragenol. Though he has done the assays (lab tests with cell cultures) that validate products for T.A. Sciences and other supplement companies, he claims not to know what the best products are! Sierra has respected corporate trade secrets and performed their analyses blind. They receive samples in a black box, identified only by a code, and send back results of the telomerase assay without ever knowing what chemical has been tested. As I write, a New Zealand company has begun to sell a skin-care product based on the most powerful known telomere drug, originally developed by Andrews and known only as TAM-818.

Andrews himself would love to develop an engineered chemical, not a natural herb, that can be optimized to have maximum effect on telomerase with minimum toxicity. He is very aware that effectiveness within the cell is only half the battle, and a successful product will have to be bioavailable in pill form, distributed efficiently through the bloodstream, and absorbed effectively into the cell nucleus. Though he claims to be very close to having the product he wants, Sierra's research has been stalled for lack of funds since their angel investors lost out to the banking devils in the massacre of 2008.

So the frustration for early adopters of telomerase supplements is that, though some products are available, there is no reliable information about which work better than others, and there is general consensus that none of them is effective enough to actually reverse age-related telomere shortening.

Instant Replay

Exercise, diet, supplements, one hormone (melatonin), two prescription drugs (metformin and selegiline), and a socially connected, self-realized program for your life. If you do all these things, you have a better shot at health and a longer life. How much longer can you expect to live? The good news: about a decade. The bad news: about a decade.

The glass-half-full crowd will say, "A decade of extra life is icing on the cake, a huge bonus for living the way I deeply want to live." Personally, I wake up every morning grateful to be a sixty-five-year-old who looks and acts and feels like a fifty-five-year-old. The glass-half-empty crowd will say, "What? Only ten years for all that trouble and bother? This isn't for me!"

The fact is that longevity beyond age ninety is increasingly a function of your genetic endowment. If you don't have the genes to be a centenarian, the recommendations in this chapter probably won't get you there. And if you do have the centenarian genetic profile, then the health and longevity program I have outlined will benefit your health more than your longevity.

For radical life extension, we must move to new technology. Fortunately, it's in the pipeline—that's what the next chapter is about. Then, in chapter 11, we'll look at some of the wider implications.

TEN

The Near Future of Aging

Breath for men to draw from as they will:
And the more they take of it, the more remains.

—Lao Tzu, *Tao Te Ching*

The Imminent Future of Antiaging Medicine

I used to write about antiaging therapies "coming in the next few years," and my biggest challenge was to convince people like you that this is not an empty promise or a far-off dream. But this year, developments in several areas are unfolding so fast that I have the opposite problem. I am sure this chapter will be outdated by the time the book gets into print, and as you read this, some of the technologies I describe should be in the headlines, and parts of this chapter will appear to be written in the wrong tense. Inevitably, there will also be ideas I describe here that will have been tried and failed by the time you read this.

Several new ideas are ripening. There are clever discoveries that have unexpectedly strong prospects, including Mayo Clinician Jan van Deursen's idea of removing senescent cells and Loyola biochemist Phong Le's discovery that a hormone called FOXN1 stimulates the thymus to regrow. There are serendipitous discoveries like C_{60} and older ideas that finally have enough foundation in understanding to become practical, including stem cell therapy, short peptides, and telomerase activators.

Most promising of all are interventions that touch the body's primary aging clocks. The perspective of this book implies that the body knows how old it is and adjusts its behavior accordingly. But how does the body keep track of time? There must be a place where the information is stored—a clock, if you will—or, more likely, several independent and redundant clocks to make sure that "if the gators don't getcha, then the skeeters will." Two known clocks are the shrinking of the thymus and the attrition of telomeres in stem cells. There are good reasons to think there is at least one more clock based on epigenetics, the timing of gene expression. We know that a child's growth, puberty, bone size, and tooth replacement are all controlled by turning genes on and off at the appropriate times. Epigenetic modulation is master of the body's development. It's a good bet that the same epigenetic clock continues to run long after adolescence has ended and that epigenetics is a central determinant of aging.

One mode of epigenetics concerns decorating the DNA with methyl groups, which can confer changes that last decades and sometimes even be heritable. In newly famous statistical surveys of survivors of the Irish potato famine, a profound tendency for obesity was found in the *grandchildren* of those starved. Epigenetics (literally, "outside-genetics") is thus involved in a kind of physiological fine-tuning of the body that is not confined to one generation. It is inherited, but not via change to the DNA sequence itself. Assays of methylation in human tissues can also be used to gauge aging, with the surprising result, for example, that breast tissue ages faster (by the methylation metric) than other organs of the body. Methylation can also linger in the body after removal of its cause. Thus quitting cigarette smoking, while slashing your risk of heart attack within weeks, will initiate benefits for methylation and cancer risk that continue to accrue over a period of years. Other processes may have beneficial effects. Many of the health benefits of exercise and caloric restriction are mediated through epigenetics, and they persist over time through methylation.

We can marvel at nature's ingenuity not only in creating varying life span strategies but in a "plasticity" that alters life span in response to current conditions. This plasticity is realized epigenetically.

There may be other clocks, as well, yet unknown to science or maybe just unknown to me.

The exciting possibility is before us that we may learn to reset the clock dial and fool the body into acting as though it is much younger than it really is. The body will take care of all the rest.

How Do These Treatments Work in Combination?

We now know of a good many substances and treatments that reliably extend life span. Of course, our first instinct is to combine them, hoping to get all these benefits at once. But the benefits don't simply add—if they did, we'd have 200-year-old humans already. How do various antiaging treatments work in combination? This is the next big question for medical gerontology.

The most accessible pathway for life extension is the energy-sensing metabolism. Many of the best-established treatments that we have now work (in part) by tricking the body into thinking it is getting less food than it really is. These include metformin, resveratrol, berberine, rapamycin, and jiaogulan. We should expect that these drugs have more benefit in people who are overweight. Moreover, they are probably mutually redundant, so that taking all of them may offer little benefit beyond taking one. Closely related are those that work on the energy metabolism, including carnosine, carnitine, CoQ10, lipoic acid, and PQQ. They may work better for people who are not exercising.

Anti-inflammatories are another class of remedies. Aspirin, ibuprofen, curcumin, boswellia, cat's claw, and fish oil work in this way, and again their benefit is predicted to be mutually redundant. If we take a few of these supplements, their benefit is probably going to max out, and taking more won't help more.

We have written about the telomeres in our stem cells as if they were an independent clock, but, in fact, this clock interacts with the others. Some of the energy and anti-inflammatory supplements have an effect on telomerase, and telomere length, in turn, affects the levels of various age signals, inflammatory signals in particular.

On the bright side, we might hope that combining treatments that work via pathways that are mutually independent should produce benefits that are better than the sum of the treatments separately. Why is that? Because the body has several alternative modes of killing itself, and if we tame any one of them, even if we eliminate it completely, the result is that the next one quickly takes over. But once we get all of them under control, the sky is the limit, and we have reason to hope that the body might thrive for a really long time. How many independent modes of aging are there? We won't know until we untangle the biochemistry a bit further, but several people in the field have speculated that the number might be manageable, say, less than ten.

We don't know this for sure. Testing of multiple treatments in mice and in humans has barely begun. There are so many combinations, so many individual factors, so many different dosages to be tried. For fifteen independent treatments, there are 105 pairs that might interact, and 455 different combinations of three. In my opinion, testing these combinations is a promising frontier for research. I have been a vocal advocate for studying combinations of treatments that have been found individually to extend life span in mammals, and I have introduced statistical methods for doing this with a minimal number of mice.

Removal of Senescent Cells

You've read this story in chapter 5 and again in chapter 9. There is a long stretch of DNA at the end of each chromosome that contains the letters TTAGGG repeated over and over and over again, thousands of times. It serves to protect the end of the chromosome from damage, and each time that a cell divides and the chromosome is copied, the copying stops before it gets all the way to the end. So chromosomes get a little shorter with each cell division.

A cell with critically short telomeres becomes senescent, not only shirking its own task but becoming destructive to surrounding tissue and the body as a whole. Senescent cells fall down on the job and don't do what they're supposed to do, but laziness is the least of their faults. In fact, they spew out signal molecules that cause inflammation, they poison nearby cells, and they are at high risk for turning cancerous—and so are other cells nearby. Worst of all, senescent cells can begin a chain reaction by inducing a senescent state in neighbor cells, even if those neighbors don't themselves have short telomeres.

Jan van Deursen is a Dutch-American biochemist at the Mayo Clinic in Minneapolis, and knowing all this about senescent cells, he wondered what would happen if you could just get rid of them. Using some tricky bioengineering, he inserted a gene into mouse embryos, creating a line of genetically modified mice in which the marker flags of a senescent cell would trigger a signal for apoptosis. A signal that a cell produces when it becomes senescent is *P16*, and Van Deursen attached to *P16* a receptor for a drug with the catchy name AP20187. The result was that cells that were senescent would take up AP20187, and it would poison them. In this way, he arranged that he could poison just the senescent cells, and this allowed a clean experiment. For these specially engineered test mice, those that got the drug would lose all their senescent cells, while those that didn't get the drug would continue to live with the senescent cells.

The experiment worked dramatically, exceeding Van Deursen's expectations. Mice were healthier after their senescent cells were cleared out. Their eyes, their bones, joints, and muscles were all rejuvenated by the treatment, and they lived 20–30 percent longer. If senescent cells were removed throughout the animals' lives, aging symptoms were delayed. If the mice were treated only later in life, there was no reversal or rejuvenation, but aging symptoms progressed more slowly.

Three years after his paper came out describing this remarkable success, Van Deursen is well on the road to being able to do the same thing for normal mice that are not genetically engineered. He has been working to develop a drug that kills just the senescent cells, perhaps one cell in ten thousand, and leave the others untouched. The drug has already proved itself in normal mice (not prepared with genetic engineering), and Van Deursen has told me that human trials are imminent.

The future of this remarkable technology is not dependent on one man. The Scripps Research Institute in California is testing dozens of compounds for the ability to turn on the cell suicide mechanism in senescent cells, while passing over normal cells. First honors go to a chemotherapy agent called *dasatinib*, and to *quercetin*, a flavonoid found in cranberries, watercress, and radishes. A handful of companies around the world have also seized the idea and are all racing to come out with effective treatments based on elimination of senescent cells. Initial applications will probably be arthritis and Alzheimer's disease, where results are remarkable and other good treatments are lacking. But the treatment promises to lower risk for all the major diseases of aging, as well, including heart disease, ischemic stroke, and cancer.

Small Molecules with Big Benefits

We met Vladimir Anisimov in the previous chapter, a veteran gerontologist at the Petrov Research Institute of Oncology in Saint Petersburg, who has also been testing longevity potions on mice through a long career. Anisimov is a glass-half-full kind of guy who has reported many positive findings, some of which remain to be reproduced in Western labs. He has done much to validate antiaging and other benefits of melatonin and metformin, but perhaps his best discovery may be epithalamin, a treatment that has been hiding in plain sight for over thirty years.

Beginning in the 1970s, Anisimov purified proteins from organs in the body that he thought were important for aging. Most proteins are giant molecules, made by chaining together hundreds or thousands of amino acids and then folding, twisting, and wrapping the chain into its characteristic shape. But Anisimov's proteins were minichains with just a few amino acids strung together. From the thymus glands of young mice, he extracted a protein he named thymalin, which seemed to offer substantial benefits. More potent yet was an extract of the epithalamus, a region of the brain that is a nexus of neural (electrical) activity and hormone (chemical) secretions. Part of the epithalamus is the pineal gland, the "third eye" from whence comes melatonin. The pineal is known as the body's day clock, in charge of cycles of waking and sleeping. There is no indication that the pineal clock keeps track of time long term, but perhaps the antiaging effects of melatonin hint at a relationship. Epithalamin is a protein of just four amino acids.

With epithalamin (also called epithalon), Anisimov was able to suppress cancer and extend the lives of rats and mice in a series of experiments over several decades. It leads to regeneration of the thymus, more potent for this purpose than thymalin. When aging humans are fed epithalamin, their mortality rates drop in half.

Why has this work attracted so little attention outside Russia? Is there anyone working to replicate it in the West or to market epithalamin? Does epithalamin (expensive and only available from foreign suppliers) offer benefits over and above melatonin (cheap and available in every drugstore)? I say it's high time the American and European gerontology communities picked up this thread.

Thymus Regrowth

The white blood cells that mount the body's most focused defense are the T cells, where *T* stands for *thymus*, a thumb-sized organ in the upper chest, just behind the sternum, where T cells are trained to recognize invaders. The thymus reaches maximum size before puberty and shrinks gradually over decades, so that it may be one-fifth its original size in a seventy-five-year-old. Loss of thymic tissue is responsible for the decreasing effectiveness of the T cells. T cells make both Type I and Type II errors. In experiments with aging mice, rejuvenating the thymus with fetal stem cells improves their immune function and improves ability to fight the most common infections. The cortex of the thymus in particular is the part of the organ responsible for assuring that no T cells that leave the thymus are dangerous to tissues of the self.

The thymus gland is a time bomb that would kill us at a certain age if nothing else got us first. It shrinks (the medical word is "involution") gradually through life, beginning in childhood and culminating in disastrous results in old age. The body's immune function is essentially dependent on a working thymus, and the immune system is related to many aspects of health, and not just by attacking invasive microbes. Our immune cells are continually attacking and eliminating errant, precancerous cells before they can become cancers. Rampant inflammation and autoimmune disorders are the consequence when the immune system begins to turn against itself late in life.

The thymus is training ground for the T cells in the blood that form the main thrust of the body's defenses. As the thymus shrinks year by year, the immune system breaks down. A ninety-year-old thymus may be one-tenth the size it was in the bloom of childhood, and this goes a long way toward explaining the vulnerability of older people to viral infections that would not be serious for a young person. Arthritis is well-characterized as an autoimmune disease, but there are autoimmune aspects of other diseases, including Alzheimer's.

There is good reason to think that if we can preserve or even regrow the shrinking thymus, then there will be benefits that echo through many or all the diseases of old age. Human growth hormone has been used with some success, but reactions to HGH vary, and there is reason to worry about its long-term effects. A recent breakthrough in treatment for the thymus looks promising enough that I have agreed to put my own body on the line in early experiments with it. A "transcription factor" is a coded chemical signal capable of switching the expression of many genes at once, turning some on and others off in one sweep. FOXN1 is a transcription factor that has been isolated from the thymus of young mice, and by reintroducing it to old mice, a Scottish research team succeeded in consistently stimulating the thymus to regrow. The larger thymus looked and functioned much like the thymus of a young mouse.

The most glaring absence in the blood as we age is of naïve T cells, cells that are not pretrained to fight any specific infection from the past but are primed to look out for new invaders. So it is most promising that the thymus glands regenerated with FOXN1 produced copious naïve T cells.

In the Scottish experiments, mice were genetically engineered with extra copies of the FOXN1 gene that could be turned on with a drug as trigger. You and I don't have these extra copies, so we need another means to get FOXN1 into our aging thymuses. FOXN1 is not something we can take in a pill, because it is a large protein molecule that would be routinely chopped up for recycling during digestion. A research group at the University of Texas is injecting little snippets of DNA (called plasmids) containing the FOXN1 gene directly into the thymus with some success. Turning on the cell's own FOXN1 gene would be ideal, and there are already candidates that can do this. There is no reason to doubt the feasibility of FOXN1 drugs, but for now, we have only rumors that they are under development.

Buckyballs

We learned in school that carbon exists in nature in two forms: graphite, black and slippery, is the common form in pencil leads; and diamond, hard and shiny, is the form that is made under high pressure and temperature, either in a lab or deep in the earth.

"Allotrope" is the five-dollar word for different chemical structures from a single chemical element. It was only in 1985 that a third allotrope of carbon was discovered, common enough to be found in soot of a dirty candle flame, but with unique properties. Sixty carbon atoms are arrayed in exactly the geometry of a soccer ball, mixed hexagons and pentagons in an arrangement reminiscent of a geodesic dome. The new compound was named buckminsterfullerene, after the inventor and promoter of geodesic architecture, who had died two years earlier.

C_{60}, or buckminsterfullerene, or "buckyball," is not a biochemical, and there is no reason to think that it would have particular biological activity. But in 2012 came a dramatic report from a French laboratory that feeding buckyballs dissolved in olive oil to rats nearly doubled their life spans. The olive oil is crucial. C_{60} molecules clump together, and it took weeks of stirring to get them dissolved in the oil. The more usual clumpy form of pure C_{60} clogs cell metabolism and is likely toxic.

The French team probably took their inspiration from a study a few years earlier at Washington University. Those researchers reported a more modest life extension of 11 percent, along with improved scores on the mouse edition of an IQ test. Eleven percent is nothing to sneeze at in a study of mammals, and if confirmed, C_{60} will join a short honor roll of substances that extend life when fed to rodents. (In contrast, there are many drugs and treatments that make fruit flies and lab worms live longer but fail the test in mammals.)

Disclaimers: there were only six rats in the experiment, and there was a glaring error in the first edition of the paper. Many researchers think the whole report may be an experimental mistake. If C_{60} does work, there are questions about how. Inside a cell, the molecule is drawn into the mitochondria, where it concentrates and scavenges free radicals. Yes, buckyballs are an antioxidant, which for me is another cause for suspicion. The Washington University team writes about C_{60} as a stand-in for one of the cell's tried-and-true antioxidant chemicals, called superoxide dismutase, or SOD.

Replication of the French report will take several more years. But there are enthusiasts who don't want to wait and are experimenting on themselves, sharing their experiences in blogs and subscriber e-mail lists.

Inflammation: The Baby and the Bathwater

Inflammation has been a recurring theme in this book, because inflammation is the most obvious and ubiquitous mode by which the older body destroys itself.

The fact that simple, "dumb" NSAIDs lower mortality in older people and increase life expectancy is very promising, but the promise is limited because inflammation has an important positive function as well as its self-destructive role. That's why the more powerful NSAIDs have side effects that limit their use. To make real progress in this area, we will need smart anti-inflammatories that selectively target the destructive role, leaving the protective function intact.

Dean Li and his research group at the University of Utah have been addressing just this challenge. Their breakthrough paper came in 2012 when they announced the discovery of a signaling pathway that controls just the destructive inflammation and is not involved in the good kind. In petri dishes, they identified a target signal called ARF6, and for therapy, they constructed a protein that contained the last twelve units at the tail end of ARF6.

Have you ever broken off half a key inside a doorknob? Not only are you unable to turn the knob, you can't pull the key out, and you can't get another key in there, either. You may have to give up on the lock and get a new one. The tail end of ARF6 works like half a key. It fits neatly into the same receptor as the full ARF6 molecule, but once inside, it doesn't change the conformation of the receptor the way that the full molecule does. It won't open the lock, and it stays stuck in the keyhole, blocking access to the real, working key.

The tail stub of ARF6 works like a broken key to interfere with the real ARF6, preventing it from doing its job. The Li lab was able to block the inflammatory reaction that responds to ARF6 without affecting the course of inflammation that is beneficial and protective.

They went on to inject their tail stub molecule intravenously in mice. They report exciting initial successes, treating mouse arthritis without gumming up the other important functions of inflammation.

Some bacteria kill the host not directly but by inducing such a violent inflammatory reaction that the patient dies of his own inflammation. Dr. Li's team challenged mice with LPS, which is the chemical that induces this fatal inflammation. Mice protected with their ARF6 tail stub had reduced inflammatory responses and mostly survived, while those without the tail stub mostly died after being poisoned with LPS.

This is a discovery that has yet to make front-page headlines, but Dr. Li's team is fully aware of the potential for changing the way we treat the inflammatory basis of arthritis and other diseases of old age, especially coronary artery disease.

Better Regulation of Apoptosis

Apoptosis is programmed cell death, and it is an essential part of a healthy metabolism. The body needs to be able to eliminate damaged and diseased cells. When a mutated cell first becomes precancerous, the cell's machinery detects the problem and induces apoptosis.

But in old age, apoptosis seems to be "dysregulated," resulting in both Type I and Type II errors—respectively, cancerous cells failing to self-destruct and healthy cells killing themselves unnecessarily. In some tissues, apoptosis appears to be on a hair trigger so that healthy cells die. One of the earliest genetic manipulations that successfully lengthened life span in mutant mice targeted the $P66^{shc}$ gene, dialing down apoptosis. (The experiment was performed with mice that were genetically altered to be cancer resistant, for fear that the failure of apoptosis would lead to too many cancer deaths. It is taken on faith that the apoptosis is optimized by natural selection for the best compromise between keeping cancer down and preserving healthy cells, but I'm not so sure.)

Some diseases of old age involve the loss of tissue, pure and simple. Sarcopenia is the wasting of muscle mass with age. "Normal" aging involves continual loss of brain cells, and in Alzheimer's disease, nerve cells in the brain are disappearing en masse. Activity of the gene *P53*, which promotes apoptosis, rises sharply in old age. Experiments suggest that cell suicide is taking away healthy, useful muscle cells in sarcopenia and nerve cells in Alzheimer's.

But apoptosis also fails progressively to protect the body against cancer as we age. Cells that would have eliminated themselves via apoptosis in a young person fail to do so in an older person. Apoptosis is also responsible for maintaining a turnover in the white blood cells that provide memory of diseases past (CD-8 cells), and as we age, apoptosis fails, leaving a residue of *anergic* (ineffective) cells that clogs the immune system.

Would it be possible to extend human life expectancy by partially blocking the apoptosis signal, reducing apoptosis generally? The question is prompted by analogy with cellular senescence and inflammation. Both of these processes take on a predominantly self-destructive role in old age, and dialing either of them down in a dumb way is an effective antiaging strategy. But making progress with apoptosis is probably not so simple. Here's a hint: calorically restricted animals actually have higher overall rates of apoptosis. My guess is that the protective role of apoptosis is too important, and so it will not be possible to extend life span by simply dialing down apoptosis. It will be necessary to restore the youthful intelligence of the system that decides which cells die and which cells live.

So then what do we do about the apoptosis problem? Not too much, at present, is the answer. Exercise helps to preserve regulation of apoptosis, in case you needed one more reason to exercise. There is a lot of evidence from cell chemistry (but not yet from whole animals) that curcumin, from the spice turmeric, potentiates apoptosis in cancer cells but not healthy cells. Ginger may have the same benefit. N-acetyl cysteine is a supplement you can find in most drugstores, and NAC has been found, like curcumin, to selectively promote apoptosis in cancer cells only. There is some evidence for similar benefits from resveratrol.

Resveratrol generated a great deal of excitement around 2003, when it was discovered to extend the life span of yeast cells. The theory was put out that resveratrol may be the ingredient in red wine responsible for the fact that the French eat lots of high-fat foods and still have low rates of heart disease. Enthusiasm grew for a few years, as worms and flies and then fish were found to live longer with resveratrol supplements. But as later studies in mice did not show a life extension benefit, enthusiasm has cooled some.

What we need are smart drugs that can be designed to selectively target apoptosis in healthy cells, which is characteristic of aging, while preserving the youthful function of apoptosis for purging diseased cells. This problem is closely analogous to the problem of inflammation described above, which similarly loses its specificity with age and becomes too active in some places while failing to act in others. In ARF6, a selective target has been identified for inflammation, but no such distinguishing target has yet been discovered for apoptosis. What is urgently needed is a way to simultaneously dial up apoptosis for senescent and precancerous cells while dialing down apoptosis in healthy muscle and nerve tissue.

The solution may come from rejuvenation of the mitochondria. Mitochondria, you may remember, are the cell's energy factories, hundreds in each cell. They also serve as the cell's executioner, triggering apoptosis on orders from the cell nucleus. It may be that the mitochondria are receiving the correct signals but not responding appropriately and that restoring youthful chemistry to the mitochondria will have the desired effect.

Your Very Own Stem Cells

You might remember from chapter 1 that our bodies have specialized cells that create new tissues. It's not true that new skin cells come from old skin cells, and new liver cells come from old liver cells. They both come from stem cells. So stem cells are the source of renewal in the body, necessary to keep replacing the cells in the skin and blood that turn over every few days and the muscle and bone cells that turn over in the course of months.

Stem cells get old and fail, and it's not because of copying errors or accumulated damage. They get old because of telomere attrition. Remember that it is cell division that causes telomeres to shorten, and stem cells are dividing over and over. In the course of a lifetime, their telomeres become short.

Telomerase activation is one way to revitalize aging stem cells. Another approach to this same problem is induced pluripotent stem cells (IPS) technology. Techniques to create induced pluripotent stem cells were discovered in 2007. A person's own readily available skin cells ("partially differentiated fibroblasts") can be reprogrammed as stem cells, grown in vitro, and then reimplanted to heal spinal injuries, ameliorate Parkinson's symptoms, treat torn tendons or cartilage or bone loss, or possibly to restore immune function lost to aging.

Ironically, it was a benighted policy of the Bush administration that spurred work on IPS technology in the 2000s. Fetal stem cells had been a staple for stem cell research, until "pro-life" activists convinced President Bush to cut off their source. The race was on to develop IPS technology as a replacement. In 2007, the laboratory of Shinya Yamanaka at Kyoto University in Japan announced the first success in turning ordinary skin cells into pluripotent stem cells. Laboratories in the United States, Germany, and the United Kingdom pushed the technology forward, making the technique more reliable, increasing the yield, and manipulating the cells toward the full capabilities of embryonic stem cells.

IPS cells are far more useful and promising than fetal stem cells, because they can be exact genetic matches for the person in whom they are implanted. Donor stem cells from a fetus are limited in therapeutic potential because they are attacked by the patient's own white blood cells, which recognize them as foreign. IPS cells can be created from tissues of the very same patient in whom they will be reimplanted, and it is expected that these cells will be recognized as "self" by the immune system, because their DNA is an exact match.

IPS technology is not yet ready for human therapy because the process is expensive, the yield is low, and IPS cells can grow into tumors. When these problems are resolved and IPS cells can be made truly indistinguishable from fetal stem cells, we will begin to learn how wide the application may be. There is a high expectation for their promise in treating Alzheimer's disease, nervous damage from strokes and injuries, damage to the heart from heart attacks, burns, juvenile diabetes, and arthritis. More speculative is the possibility that skin, bones, blood, and other major systems of the body will be made more youthful by a fresh supply of stem cells. The immune system weakens with age, in part, because blood stem cells are not churning out new white cells the way they do in a young person.

No one knows why or how, but stem cells have a remarkable ability to go where they are needed and to grow into whatever tissue is required locally. Where does this intelligence come from, and how do they navigate? I don't think we even have a theory. But this ability makes the application of IPS useful and practical in a way it might never be otherwise. We will learn in the next few years whether infusions of stem cells can confer a general antiaging benefit on the body.

Telomeres! Are They the Body's Primary Aging Clock?

In the short run, removing senescent cells looks like a promising technology. But wouldn't it be even better if we could maintain telomere length so cells don't become senescent in the first place?

The basic story of telomere loss and cellular senescence was known in the 1990s due to the pioneering work for which Elizabeth Blackburn and Carol Greider won the 2009 Nobel Prize. But it was widely assumed that telomeres must be irrelevant to human aging. No way could telomere length limit human life span, because the remedy to telomere aging was right there at hand in the form of telomerase, the enzyme that restores telomere length. Every cell knows how to make telomerase; it's easy, and there is virtually no energetic or metabolic cost. But in humans, as in most mammals, telomerase is only expressed in early stages of embryonic development. Cells of the embryo grow telomeres that are long enough to last a lifetime. The usual assumption is that the body knows what it is doing and is out to protect Numero Uno. So if the cells are not expressing the telomerase gene during adulthood, it must be that it is not needed. There must be some benefit from sequestering telomerase where it is hard to reach. Maybe keeping telomerase out of easy reach makes it more difficult for a cell to turn cancerous and replicate out of control . . .

This way of thinking blew up in 2003 when a powerful link was discovered between life expectancy and the length of a person's telomeres. We introduced Richard Cawthon in a box on page 164. He pioneered some of the technology for measuring telomere length and used the new technique to analyze archived blood samples from people who had donated blood twenty years earlier. He then followed up to ask about the health and mortality of those people during the intervening years. Essentially, he dug up a prediction from the past and found that it foretold who was to die and who was to live. People with shorter telomeres had far higher death rates from infectious disease and especially from heart attacks. They did not have lower cancer rates than people with longer telomeres.

In 2001, the technology for analyzing DNA sequences was newly available, though monumentally expensive by today's standards. The race to sequence the human genome had ended in a dead heat, the renegade private entrepreneur Craig Venter up against the full resources of the National Institutes of Health. With the new technology, one could measure telomeres in a sample of people to see whether there was any relationship to health outcomes. As some fell sick over the coming years while others remained healthy, the data would trickle in, and an answer would gradually come into focus. But Cawthon leapfrogged over this slow course by obtaining blood samples from people from decades past and their medical records for the intervening years. A professor of biochemistry at the University of Utah in Salt Lake City, with access to the data of the close-knit Mormon community that tended to stay put, Cawthon obtained 143 archival, frozen samples of blood from the University Hospital. All subjects had been about sixty years old at the time the blood was drawn. He traced what had happened to each of these people over the intervening years—who had died, who had lived, who had been healthy, and who had fallen sick. The correlation he found was unexpectedly tight, so strong that even with his tiny sample, you couldn't miss the correlation. People with the shortest telomeres were dying twice as fast as people (the same age) with the longest telomeres,

Do short telomeres really cause diseases, or are they a passive record of a person's history of disease? Cawthon established an association between telomere length and mortality that has stood up in many replications since then, both in animals and in humans. But correlation is not the same as causality. There is an alternative hypothesis that short telomeres represent a kind of scar from past stress, when cells were required to replicate more frequently to recover from infection or injury. Twelve years after Cawthon, a Danish group repeated his study with a huge sample of over 65,000 people. They confirmed Cawthon's findings, and added valuable detail. The Danish study added weight to the hypothesis of a causal link between short telomeres and higher rates of cancer, heart disease, and death.

The hypothesis that telomeres may be one of the master clocks controlling the rate of aging in our bodies was considered radical when Michael Fossel first promoted it in the late 1990s in a book, *Reversing Human Aging*, that now seems prescient. Cawthon and follow-up studies did a great deal to make this hypothesis plausible at the same time that another popular book by another Michael, *The Immortal Cell* by Michael D. West, espoused the thesis that telomere attrition is an important cause of aging. This was a period in which it was becoming increasingly clear that there were mechanisms by which cells with short telomeres could cause some of the diseases of aging: by becoming senescent and poisoning other cells nearby, and dialing up the inflammation that is one of the primary ways in which the aging body destroys itself.

It may be that telomeres are able to exert their influence even before they become critically short. Even though the telomere represents only a tiny fraction of the length of a chromosome (less than 0.01 percent), telomere length nevertheless affects the way the whole chromosome folds and which parts of it are open and available for gene expression, and which parts remain under wraps, silent and unavailable. The University of Texas laboratory of Woody Wright (my classmate at Harvard decades ago) claims to see thousands of genes whose expression is affected by the length of the telomere.

The idea that telomeres may be an important cause of aging fits handily within the thesis of this book. For aging to be programmed on a schedule, the body needs a clock (or several) to keep track of how old it is. Telomere length makes a handy clock, and it is readily modulated with small amounts of telomerase that are known to be expressed during adulthood. The toxicity of senescent cells provides a convenient suicide mechanism, and a small number of senescent cells is sufficient to do the job.

It may be that one of the ways the body shuts itself down is simply by rationing telomerase and letting telomeres run short in the stem cells that are responsible for renewing our tissues. This implies that lengthening telomeres may be a powerful intervention, a unique opportunity to reset an aging clock.

Most gerontologists think that it cannot be so simple. Some say that lengthening telomeres would not be effective, and others fear that it would actually cause cancer. But people who are most prominent in telomere research tend to believe that telomerase will prove to be the philosopher's stone, the fountain of youth, the elixir of Gilgamesh about which humanity has dreamed for thousands of years.

What Can We Expect from Telomerase Activators?

In 2003, Richard Cawthon discovered that shorter telomeres were related to higher mortality in a study of just 143 people. By 2015, a Danish group replicated his work in a study that included 65,000 people. Cawthorn's subjects were all about sixty years old, but the Danish study encompassed a wide age range.

With a sample so large, it is possible to see the effect of telomere length on health and separate it from the effect of age. By comparing the life expectancy of people with short telomeres and people the same age with long telomeres, it is possible to make a guess what the effect would be if the people with short telomeres could be treated so as to regrow their telomeres to a fully youthful state. This method yields the answer: four years of extra life. For many scientists enthusiastic about telomere therapy, this result is disappointingly small. For those who think that the idea of an aging clock in our telomeres is preposterous, four years seems absurdly large.

As discussed in the previous chapter, simply taking a telomerase pill is not going to work. Telomerase is a huge molecule that will be broken down in the digestive tract, and none of it will reach into the cell nuclei where it is needed. Injection of intravenous telomerase has never been tried, even in lab mice. The thinking is that, even after bypassing the digestive system, too little of the telomerase would find its way into stem cell nuclei.

There are now many companies working on antiaging interventions with telomerase. Some have tried to deliver telomerase in a form that will reach the cell nuclei, but a more common and promising approach is to find substances—drugs or herbs or supplements—that signal the cell to switch the telomerase gene *on*. The aim is to stimulate the body's native telomerase expression so that telomerase will appear right there in the cell nucleus where it is most needed. Extracts from the Chinese herb astragalus seem able to induce telomerase in the nucleus, but it requires a mountain of astragalus to obtain an effective dose. Perhaps the active chemical constituent can be synthesized. T.A. Sciences New York has been selling the astragalus extract to early adopters since 2007, and they report evidence of modest success in slowing the rate of telomere shortening. They sell their formula under the brand name TA-65, but the proprietary formula is rumored to be a highly purified extract of the astragalus root, cycloastragenol. The product appears to be effective but perhaps too weak to be practical. In tests with mice, the dosage corresponds to a human equivalent of 1500 mg/day. But capsules that are sold currently (2014) contain only 5–20 mg and still cost several dollars per pill. What we need is a cheaper and more effective alternative.

The amino acid carnosine, which is sold as a nutritional supplement, may stimulate the body to express telomerase. Omega-3 fatty acids (from fish oil) have also been suggested as an aid in maintaining telomere length. Herbs called ashwagandha, bacopa, and milk thistle have some activity, as do the spice curcumin and the red wine elixir resveratrol. Stress management and even meditation may have measurable effects. But all these together cannot keep up with telomere loss over time. The breakthrough technology that will safely and cheaply maintain our telomere length has yet to be discovered. Sierra Sciences, Bill Andrews's company in Reno, leads the field, after screening hundreds of thousands of molecules in cell cultures for the ability to induce telomerase expression.

Meanwhile, experiments continue to taunt us with the promise of rejuvenation through telomerase. Mice in a Harvard lab were genetically engineered with a chemical on/off switch for telomerase. Without it, their bodies showed severe symptoms of accelerated aging, their brains atrophied, and their senses suffered. When telomerase was turned on, these mice were rejuvenated in a number of ways—most impressively, in regrowth of the brain and restoration of their lost senses. In the Spanish laboratory of Maria Blasco, the modest potency of TA-65 was demonstrated to have measurable effects in mice. In another experiment, the same lab found that mice lived longer when they used genetically engineered viruses to insert an extra copy of the telomerase gene in living adult mice. This is both encouraging and puzzling, since lab mice (unlike people) are thought to have so much telomerase that their aging must be independent of telomere loss. American entrepreneurs have begun to offer offshore clinics at which people can pay to be among the first humans to receive gene therapy with telomerase.

A frustration for everyone in this field (and for eager consumers on the sidelines) is that the for-profit companies that are funding research have cloaked their findings in secrecy. Large pharmaceutical companies routinely keep their research projects under wraps. I have personally quizzed several top telomere scientists about the state of the art in telomerase activators, and I am convinced that they are sincere when they say they don't know. The Internet buzzes with rumors about wildcat treatments for humans in the unregulated environment of China. T.A. Sciences won't tell us what their flagship product is made of, and Isagenix lists dozens of ingredients in their Product B, some of which have nothing to do with telomeres. Sierra Sciences' blind tests for other companies of the effectiveness of hundreds of herbal supplements, chemicals that come to them in a black box with just a code, shows that the race to extend the life span of the human race is on.

Heterochronic Parabiosis: The Vampire Cure

In the nineteenth century when surgery was new, doctors discovered that they could join two animals together, artificial conjoined twins, in which the veins and arteries at the juncture would find each other and grow together so the two animals would share a common pool of blood. Previously, the art and science of grafting plants had been pursued for centuries. Following the lead of Polish entomologist Stefan Kopeć, Vincent Wigglesworth in the 1930s joined beheaded insect larvae neck to neck via paraffin wax, showing that a hormone transferable from one insect to another was responsible for molting. Wigglesworth called the creepy body joining "parabiosis," and "heterochronic" simply means that the conjoined animals are different ages. These experiments were resurrected in the 1950s by Clive McCay, the very same McCay who discovered the life-extending effects of caloric restriction two decades earlier. It was probably the unsavoriness of these experiments that prevented their full potential from being realized for many years.

In the early 2000s, a young couple fresh from Harvard joined Tom Rando's Stanford lab as new grad students, and parabiosis experiments between young and old mice were married to modern biochemical analysis. Maybe the Russians are less squeamish, I don't know, but the enthusiasm and expertise of Mike and Irina Conboy was the source of the current wave of research. In the first study from the Conboys and other members of Rando's lab, muscles of old mice were shown to heal better when nurtured with the blood of a younger mouse. The specialized stem cells showed signs of rejuvenation. Old stem cells were newly empowered, acting like young stem cells.

Since 2005, the authors of this study have graduated and founded their own labs at Berkeley and Harvard. Together and separately, they have made steady progress on the road from the witches' cauldron of parabiosis to practical treatments for the diseases of old age.

The first step was to discover that the rejuvenation has nothing to do with blood cells, red or white! Instead, it comes from dissolved proteins and RNAs in the blood plasma. These are hormones and other chemicals, secreted and circulated through the body, signals that keep various parts of the body on the same page. Among them are transcription factors that find their way to the cell nucleus, latch onto the chromosomes, and switch whole suites of genes (on or off) in one fell swoop.

The next step was to show that blood plasma from a young animal can rejuvenate an old animal. Tony Wyss-Coray at Stanford has done this with mice that are genetically engineered to get Alzheimer's disease. Intravenous plasma transfusions are much less invasive than conjoining animals in parabiosis, but some of the benefit is short-lived, and we don't yet know how frequently the procedure needs to be repeated. Blood plasma from young mice stimulates nerve regrowth, and the mice recover some of their memories and other brain functions. Wyss-Coray has leapfrogged over many phases of trials and validations to try human plasma transfusions in patients who are at a late stage of Alzheimer's disease. But everyone agrees that this is not a solution for the long term. It may be that old people can be made young with repeated transfusions of young plasma, but it is an expensive, time-consuming procedure that involves many young blood donors for every old patient. This is not a practical rejuvenation program for the masses.

Ultimately, what we would like to know is, what substances in young blood are responsible for the rejuvenating effect? Equally important, there are signal chemicals in old blood that need to be removed or blocked, because they herald self-destruction. If we are lucky, there will be a manageable number of such signals that can be isolated and synthesized. Subtracting blood factors need not be more difficult than adding them. The technology was described a few pages back as the "broken key" technique. We may hope that a small number of powerful transcription factors can be rebalanced in the blood of an old person, and the old body will respond to the signals, becoming a young body.

Irina Conboy has told me she has a hunch what some of these blood factors are, both those that need to be added and those that need to be removed. Tony Wyss-Coray, Amy Wagers, Saul Villeda, and other people doing this research probably have their own hunches. In 2014, Wagers wrote about impressive rejuvenating effects from a little-known transcription factor called GDF11, and Conboy announced powerful benefits from the well-known hormone oxytocin. Oxytocin is activated during childbirth, breast-feeding, and acts of love, and Conboy discovered that the same chemical promotes muscle regrowth and healing. GDF11 is a growth differentiation factor that stimulates regrowth of nerves and blood vessels, in addition to muscles. An energy enzyme called AMP kinase is also lost with age and may be another candidate for a boost. Already, doubts are being expressed about GDF11, because it is a form of transforming growth factor beta (TGF-ß), and it may actually do more inhibiting than promoting of growth. The debate is robust, and we expect that the process will home in on the blood factors that are the upstream source of aging within a few years.

NF-κB is an inflammatory signal that will probably have to be removed from the blood. Other signals that we have too much of as we age include luteinizing hormone (LH) and follicle-stimulating hormone (FSH). LH and FSH are hormones involved in the female reproductive cycle that perversely come out in force after menopause with corrosive, pro-aging effects. Cortisol is a stress hormone that is overexpressed in old age; it is the primary poison that kills salmon after they are done with their reproductive tasks. TGF-ß is a signal that dials up apoptosis, causing healthy and useful cells to eliminate themselves with effects that are gradually crippling.

Some of these signals are primary and control whole cascades of other signals. Basic research will untangle their mutual relationships, and with luck, we will discover a few master transcription factors that can be rebalanced with powerful rejuvenating effects. If we are not so lucky, it may turn out that there are hundreds of signals, all intertwined and controlling the body's age state "democratically," in the sense that there is no hierarchy with a few on top controlling many on the bottom. In this case, rejuvenation through blood factors will turn out to be a much more difficult task.

There's only one way to find out.

Alexander Bogdanov was a Russian physician, utopian Bolshevik, and visionary novelist. When he reached fifty, Bogdanov was troubled by symptoms of his aging body, and he began a series of experiments with blood transfusions from young donors on himself and others, including Lenin's sister. "After undergoing 11 blood transfusions, he remarked with satisfaction on the improvement of his eyesight, suspension of balding, and other positive symptoms. The fellow revolutionary Leonid Krasin wrote to his wife that 'Bogdanov seems to have become 7, no, 10 years younger after the operation.'" But his eleventh transfusion came from a young student who had malaria, and Bogdanov fell ill and died at the age of fifty-five.

As of this writing, the technology of rejuvenation through blood transfusions is promising, though many questions remain about the basic technology. But already, an ambitious med-tech start-up has jumped the gun and is offering transfusions in offshore clinics for hoary tycoons.

The Epigenetic Aging Clock

The paradigm of programmed aging implies that the body deconstructs on a schedule. So, we may ask, what controls the schedule? There must be some master clock, or several. These biochemical clocks will be the prime targets for high-leverage intervention that will restore youthful function and prevent many diseases at once by tricking the body into believing that it is younger than it actually is.

We have identified two aging clocks that operate over a lifetime: loss of telomeres in the stem cells, leading to cell senescence, is one; involution of the thymus, leading to collapse of the immune system, is the other.

But there are good reasons to believe that there are other aging clocks. For one thing, the thymus and immune systems would seem to be connected to some, but not all, aspects of aging. Aging of the brain and the skin do not have any obvious connection to the thymus. For another, there are animals that age and die though they have plenty of telomerase through their lives. Bats and pigs and laboratory mice, for example, do not lose telomere length as they age.

Where is this clock that we cannot see, and how does it work? Recently, several scientists in this field have proposed that epigenetics constitutes an independent aging clock.

Our development and maturation unfold systematically, reliably on a schedule. Girls begin to menstruate on average when they are twelve; boys grow hair on their chests at about sixteen, and both get four new teeth in the backs of their jaws around eighteen. These events are not what we would call aging, but they are timed and programmed by changing gene expression. What causes the gene expression to change?

I like the idea that gene expression constitutes a clock in its own right, perhaps the most basic clock of them all. Gene expression is controlled by the epigenetic state of the chromosomes, and this is truly a central command, telling the body what to do. Among other things, the epigenetic state controls what hormones are circulating in the blood, including those transcription factors that control and program the epigenetic state. The epigenetic state of cells all over the body is part of a feedback loop that could be the basis of a clock. Today's epigenetic state determines which hormones circulate in the blood, and some blood factors come back into cells, entering the nucleus and programing the epigenetic state for tomorrow. This is the idea behind the epigenetic clock. It would be a single, unified clock in charge of development in the embryo, growth in the child, maturity in the adolescent, and aging in the adult; apoptosis, telomerase rationing, and progressive immune dysfunction could then have been secondarily recruited as optional pieces of the aging program. All this activity is orchestrated by varying the composition and quantities of hormones, RNAs, and other factors in the blood plasma.

Steve Horvath is a computer scientist at UCLA who has studied the relationship of aging to methylation of the body's DNA, using ideas that are purely statistical and blind to biology. He has found combinations of hundreds of methylation markers that collectively report the age of a person with remarkable precision. I suspect that he may be close to a fundamental cause of aging. I suspect there is a feedback loop in which gene expression controls the state of the metabolism, and then the state of the metabolism affects gene expression. This is the basis for a biological clock. Perhaps the entire time sequence for growth, development, and aging is dictated from neuroendocrine glands in the brain.

The epigenetic clock hypothesis is closely related to the parabiosis experiments and the search for blood factors that control aging. Early success of these experiments suggests that we may indeed be lucky, that an optimistic scenario described will obtain, and that just a few chemicals will have the power to change the age state of the body.

The epigenetic clock concept may also dovetail with the telomere theory of aging. Telomerase may have powers as a transcription factor, in addition to its well-known role growing telomeres. And telomere length itself is known to have an influence on the expression of many genes, so telomeres are logically a part of the epigenetic state.

My guess is that the unifying concept of the epigenetic clock is our best guiding paradigm for research in antiaging medicine. From a broad view of the biology of development and aging, it looks plausible. And in any scientific search, it makes sense to try the easy path first; the prospect of reprogramming an epigenetic clock lights the way to a fast track in the landscape of rejuvenation science.

Instant Replay

Inventors are very practical people—they have to be. Medical researchers have realized that, compared to decades of plodding research on cancer, heart disease, and Alzheimer's, the prospect of even incremental changes in human aging have enormous repercussions. For example, setting back the aging clock by just four years would save more lives than a perfect cure for cancer. Like a battleship in midcourse, government research priorities are slowly shifting to accommodate this reality, while some individual researchers and biotech companies are oceans ahead of them.

Some researchers are purely practical, doing what works. They are intervening to thwart the body's self-destruction machinery without worrying whether their intervention is "natural" or thinking about the evolutionary origin of that machinery. For example, killing senescent cells. Jan van Deursen is a biochemist par excellence; he is not a philosopher or even an evolutionary theorist. He does not ask, "Why has the body put up with cells that are murdering it?" He focuses on the task at hand—first proving that eliminating these cells extends life and then solving the biochemical puzzle of how to tag them selectively and eliminate them without harm to innocent bystanders. Mike and Irina Conboy are agnostic on the question of programmed aging, but they are at the forefront of research to detect the essential differences between signal molecules in the blood of old people and young people.

Some other researchers of aging have already absorbed the message of this book and have focused their lasers on one or another of the body's aging clocks. Bill Andrews and Michael Fossel both understand that telomeres shorten explicitly in order to kill us, and they are (separately) pursuing the most advanced and credible programs to reverse aging with telomerase. Greg Fahy recognized two decades ago that the shrinking thymus is the immune system's aging clock, and he has spent those decades laying the groundwork for clinical trials of a credible scheme to regrow the thymus.

There also remain many researchers who are being held back by evolutionary dogma and don't even know it. The worst example is that the potential of telomerase therapy is being held hostage to a pernicious myth that telomerase must cause cancer, a pervasive belief that now has a life of its own with a circle of experts citing each other's authority despite the complete lack of supporting evidence.

Liberated from the ideology that says that the body is already doing its best to live as long as possible, that insists "it cannot be so simple" and that "there is no free lunch"—new avenues for research will open up. The future of antiaging medicine looks bright. The largest cloud on the horizon is the challenge of an epigenetic aging clock, coded in a language we have yet to crack.

ELEVEN

All Tomorrow's Parties

The majority, though they know not what to do with this life, long for another without end.

—Anatole France, 1921 Nobel Prize in literature

That tadpoles are fodder for pond-life is as natural as the leaves falling on the water in autumn; that flies get squidged is as ordinary as apples rotting in the orchard. One's own death, on the other hand, seems most unnatural. It seems rather an error and an outrage; a cosmic crime; a reason to raise one's fist and rebel against the regime that ordered this slaughter of innocents . . . Nature is a streamers-and-all, non-stop, cork-popping party of death . . . but that does not mean that death is right or good . . . The death of any creature, even a fly, is a catastrophe; but at the very same time, from the viewpoint of the gods, the deaths of us and the flies are equal in their insignificance.

—Stephen Cave

Puzzles and Paradoxes

This book has been built around two major themes:

- Natural selection has crafted a death program in our genes, because limitless life spans lead to unsustainable population expansion, followed by ecosystem collapse, risking extinction.
- With some understanding and ingenuity, clever humans can defeat nature's death program and gain for ourselves much longer and healthier lives.

The astute reader may have apprehended that bullet #2 is in flat-out contradiction with bullet #1. And some of you softhearted, liberal collectivists might even entertain doubts about the wisdom of conniving to overstay our invited three-score-and-ten years on this finite Eden, which we hold in trust for our children's children.

Truth be told, maybe we two who are your hosts for this final and most speculative chapter share some of these apprehensions. Our wildly breeding population, whose planetary presence just a thousand years ago could fit into what is now the state of Massachusetts, is playing a dangerous game. Despite our big brains, our technology, and our hopes for a better tomorrow, we have not yet shown a level of collective responsibility to match the intelligence that is evolved in our genes. Having received the gift of senescence from our ancestors, scared mammals evolving better vision and bigger brains as they ventured into a post-Cretaceous daylight free of predatory reptiles, we have gone on to make a mess of things.

But sanctimonious admonitions do not sell books, and so our editors at Flatiron Books have enjoined us not to touch this topic with a ten-foot pole. The current chapter has been slipped into the manuscript when they were not looking.

Levity aside, we acknowledge this contradiction in all its immensity and report that we have wrestled with it as the most challenging issue connected with antiaging. More broadly, this paradox must haunt anyone who aspires to improve the human condition. We recognize

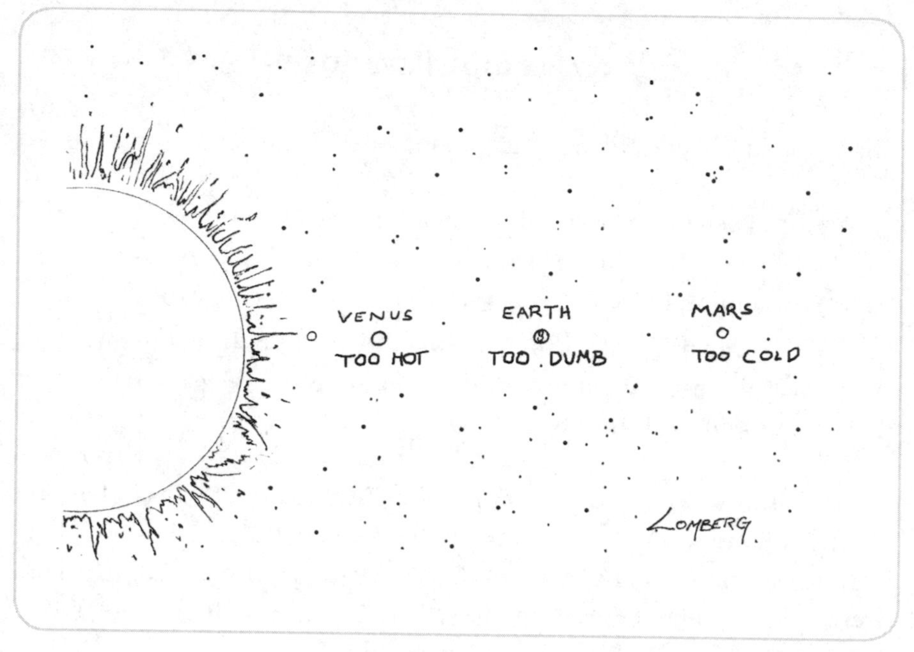

Fig 3. *Intelligent Life? (credit: Jon Lomberg)*

the disastrous consequences of man's ascent to become Gaia's new top predator. The entire planet is being transformed into a monoculture for the human parasite, which already has spread to the poles and seven seas and threatens to expand into space.

On the one hand, we feel empathy for individual humans and wish sincerely to help them to thrive. On the other hand, the potential for conflict between the welfare of the individual and the welfare of the community is deeply ingrained in our thinking; and we are well aware that when "the community" collapses, there are lots of individuals who perish with it.

We observe with requisite horror that the collective consequence of too many humans thriving for too long has been a narrowing of life's diversity and lush potential, a devastating triumph of the Shopping Mall over the Rain Forest. To these and other existential dilemmas, we have, alas, no answers. But this chapter is the tribute we pay to these great and open questions.

The Social Implications of Longevity

The nursing home is our worst fear about life extension. We imagine what would happen if the life extension we achieve were only "at the end," staving off death with one more bypass operation or one more round of chemotherapy. We would pass those extra years in long-term care, unable to function or to enjoy life.

Some social demographers fret about the increase in life span, because a smaller working-age population will end up supporting a larger population in nursing homes and hospitals. They have it exactly wrong. In fact, all the technologies for extending our life span increase health span, as well. This means that we can expect longer careers, more productivity, and—most relevant—an improved ratio of healthy years to years of disability. We will get used to retiring at seventy-five or eighty-five or ninety-five. We will accumulate more experience and maintain our productivity longer as we retain the health of our brains as well as our bodies. In the best scenarios, we will become a race of sages.

The fear that the part of our life span that we can readily extend will be the painful, disabled part is unworthy of credence. It is, and will always be, far easier to keep a healthy person healthy than it is to keep a failing body on life support. Likewise, the idea that there won't be enough young workers to support a top-heavy population residing in nursing homes is a red herring.

But increasing longevity is connected to other problems that are all too real.

In a classical Greek myth, Eos, Titan of the Dawn, fell in love with the boy Tithonus, beautiful but mortal. Smitten, she despaired that he must one day die, leaving her forever without him. She prevailed upon Zeus to intervene, to give Tithonus immortality. Zeus complied. But Eos forgot to ask also for Tithonus the eternal youth that was part of the package for the gods. So Tithonus was cursed with the worst of both worlds: he was trapped forever in a body that became ever sicker, more painful, and decrepit, but unable to escape into death.

Does Life Extension Contribute to Overpopulation?

Well, yes, of course it does. Before 1800, the human life expectancy had been stable for thousands of years, and the world's population had been kept in check. But in the past two centuries, the average human life span has increased at the rate of one added year of life for each four years of history. The current explosion in human population has been going on for two hundred years, fueled not by higher birth rates but by lower death rates. In fact, a wave of increasing life spans has accompanied the advance of hygiene and medical technology, and a wave of decreasing birth rates has followed close behind. It is the time lag between declining deaths and declining births that is responsible for population growth. Currently, Africa is the last continent where technology is finally moving in to increase life expectancy, and the African birth rate is coming down, but not fast enough to avoid devastating population increases. Aldous Huxley recognized this pattern as early as 1956. "What we've done is 'death control' without balancing this with birth control at the other end . . ."

A demographic revolution has paralleled the industrial one. Life span in 1840 was about forty years in the world-leading European countries. In 2014, it was eighty-three in Japan and Scandinavia, the present world leaders. And indeed, the increase has been quite steady and gradual, so that "one year for every four" is an accurate characterization.

For the first 120 years, increased life span was a story of early deaths prevented. Before about 1970, all this progress in life expectancy was achieved by protecting people from dying young, benefits reaped from antibiotics, hygiene, and workplace safety. Many lived to seventy, but the proportion who lived into their eighties and nineties remained small. It was widely expected that life expectancy would level off somewhere in the vicinity of seventy years, and further advances would be very slow, if they happened at all.

But since 1970, a remarkable thing has happened: the maximum human life span has risen, and it continues to rise at an accelerating pace (as recounted by James Vaupel). What is more, people in their seventies and eighties are healthier today than ever before in human history. Even though there are dramatically more seniors in the population, the proportion of the population in assisted living and other dependent care situations is not rising. This is just what we wanted—we are staying active and healthy longer, retiring later, delaying the ravages of old age, and "compressing morbidity" of late life into a shorter endgame.

In the early stages of this sweeping increase in life span, the number of people of childbearing age increased steadily. This compounded the explosive effect on population. Since 1970, the increase in life expectancy has been focused past the age of childbearing, and so the effect on population has been far more modest. Still, it goes in the wrong direction and contributes to the cancer of overpopulation.

Are We Destructive to Earth's Ecology?

The good news is that the Earth is still green. The bad news is that we the people have already arrogated for ourselves the most fecund and productive regions of the planet, disrupting the habitat where most of Earth's animals and plants make their homes. We are in the midst of the sixth great extinction event in the history of macroscopic life.

The most famous extinction event is the one that ended the reign of the dinosaurs sixty-five million years ago. About 30 percent of all species disappeared after a meteor, estimated to be ten kilometers across, crashed into the Gulf of Mexico with an energy comparable to all the nuclear weapons now in existence. Dust clouds darkened the sky for several years.

There have been four other mass extinctions in the 500 million years that animals have roamed Earth. The largest of these, called the Permo-Triassic, or PT extinction event, took place at the halfway point, about 250 million years ago, and its cause is still debated. Fifty percent of all species perished, including a huge proportion of marine species. Vertebrates had been around for most of that first 250 million years, but it is conjectured they got a big break after the great PT extinction.

To be sure, early human civilizations had been responsible for some spectacular extinctions, from the saber-toothed cat to the woolly mammoth to the Australian moa. But the sixth extinction, described in Elizabeth Kolbert's sober and readable book of the same name, began in earnest in the twentieth century. The ecology of the oceans is unrecognizable compared to what it was one hundred years ago, driven by mankind's preference for larger fish. Vast drift nets have been devastatingly effective in denuding most of our oceans of all large predators. Removing the predators at the apex of an ecosystem has an effect that ripples through and disrupts the balance at every level, down to the plankton and the coral.

On land, as well, species are being lost at a rate that dwarfs the ability of scientists to catalog them. There is an accelerating domino effect, as loss of a keystone species makes it inevitable that an entire ecosystem will succumb over the ensuing decades. But exactly how rapidly species are disappearing is a subject for vigorous debate. Estimates range from a low of 0.01 percent per year (from right-wing think tanks) to E. O. Wilson's well-reasoned, alarming prediction that half of all species will vanish over the next century, victims of an Anthropocene ecocide.

Will We Destroy All Life on Earth?

A reincarnated Jonathan Swift might be tempted to portray us as tiny antlike scientists thinking we are going to make a difference to a spherical behemoth with considerably more experience.

Don't give yourself airs.

Eradicate Gaia? Life is bigger and more robust than anything we are able to disrupt. No, the threat is not to Gaia but to ourselves. Life will eventually roar back, more diverse, more wondrously inventive than ever. But recovery from a mass extinction requires, typically, a few tens of millions of years. That's nothing for Gaia, but for our grandchildren, thirty million years may try their patience.

There is life in boiling-hot sulfur pits and life on the pitch-black ocean floor, thriving under pressure that would crush a scuba tank, and life embedded in dry rock, equally deep under the land, living on who knows what. There are spores that were trapped in salt deposits two hundred million years ago, recovered by scientists and brought back to life in the laboratory. To eliminate all life on Earth is far beyond humanity's destructive power for the foreseeable future.

But can we imperil the ecosystem that sustains human life? Quite possibly we can.

Suppose we were to turn on a dime, adopt the ethic of Chief Seattle and become stewards of the biosphere. James Lovelock, originator of the Gaia Hypothesis and one of the first to sound the alarm on the catastrophic dangers of global climate change, suggests that we may not be fit stewards of anything. We have made a mess of things and had best focus on ways to preserve ourselves, rather than thinking we can care for, let alone "save" the entire planet. For rampantly reproducing humans to imagine themselves responsible for stewarding this beautiful blue marble makes as much sense as tasking goats with managing the garden or putting foxes in charge of the proverbial chicken coop.

"Thank God, men cannot as yet fly, and lay waste the sky as well as the earth!" wrote Henry David Thoreau in his *Journal*, half a century before the Wright brothers. "Most men, it seems to me, do not care for Nature and would sell their share in all her beauty, as long as they may live, for a stated sum—many for a glass of rum."

Nevertheless, the consciousness that alerts us to our own impending nonexistence (as individuals) informs us similarly of a possible fate for our populations and kind, and part of us, having escaped the magical circle of self-justification and denial, acknowledges a still, small voice, counseling that we are all in this together.

One of the great correlates of the present book is that nature is not only (as Tennyson wrote) "red in tooth and claw," or (as Hobbes put it) a *bellum omnium contra omnes* ("war of all against all"). The living world is also a vast symbiotic collective of groups and constantly shifting, multilevel alliances in which the most cooperative partnerships and conglomerates triumph. Such triumph can be short-lived, even Pyrrhic.

Historically, putting the brakes on maximal growth and reproduction has involved killing, of both cells and individuals. This is the face of the Black Queen, and though we may regard her as a monster, she is intimately connected with our survival as a species among species. And this crucial observation has been elided—missed, forgotten, played down, and plain ignored by popular expositors of evolutionary theory, with major consequences for human ecology and human health.

Are We Special?

We are senescing animals, gifted with exquisite perceptions and sensation, able to see and hear and fear and suffer and feel joy and perch atop ecosystems where plants do the food production, and fungi and bacteria do the cleanup and recycling. Religious fundamentalists flatter themselves to think that an all-powerful God arranged this little planet for our profit and delectation. But scientific humanists and transhumanists are not immune from this conceit. Ecological science suggests we are not so special. And part of our non-specialness may be our vulnerability to the consequences of excessive growth from which aging has evolved to protect us. Our intelligence evolved to promote our survival and growth when these words were synonymous, and now we must see whether the same intelligence can serve us to temper growth for the sake of survival. For billions of years of life's history, the multifaceted genetic program we've called the Black Queen appears to have been tempering population growth. This program was shaped by a violent process that must have involved many mistakes, many extinctions. Consciousness and technology, like sex and the Black Queen, speed up the rate of evolution but may not insulate us from the pain of failure, the great die-offs that have been part of nature's toolkit through the aeons. Repeated failure provides the working medium from which nature carves her brilliant sculptures. Are our brains so much smarter than our genes that they might learn the same lesson without first suffering catastrophic loss?

The geological era that started at the close of the last Ice Age, continuing into the time in which we were born, has been named the Holocene era, but the accelerating influence of human numbers and human technology on life everywhere has led geologists to speak of an Anthropocene era. Donna Haraway of UC–Santa Barbara argues that, more accurately, the culprit is one particular form of human organization, and so a better term might be "Capitalocene." In her work, she explores the importance of interspecies relationships, of community, pleasure, and responsibility for the lives we have altered as we move into reformed ecosystems that might be stable in the long term. Ultimately, she concludes, what is required is a return to slower growth and natural living characteristic of pre-patriarchal societies and, indeed, of life itself. Playfully alluding to an H. P. Lovecraft story, she suggests we need to move from Capitalocene to the "Chthulucene"—from the same root as "chthonic," referring not to humans but to our supporting living environment, a.k.a. Earth.

David Sloan Wilson (my mentor, you may remember from the introduction), has written that many life-forms cooperate more reliably than do humans, and he cites shrimp, coral colonies, and naked mole rats as well as the familiar bees and ants as examples. Democracy is what distinguishes human cooperation. Other deeply social species have colonies of sterile workers who sacrifice their genetic legacy for the sake of a single queen who reproduces full time. The uniqueness of humans lies in our ability to negotiate terms of cooperation that are fair and evenhanded. The motivation of human workers compares to the motivation of a worker ant as an ESOP employee compares to a slave. The freest and most democratic of human societies, argues Wilson, have proven to be the most formidable competitors in the history of the biosphere.

We agree, in principle, and note that democracy is waning in the Western world, led by the shameful example of the United States. Representative government has degenerated into a façade for a vast, authoritarian bureaucracy that does the bidding of the world's most powerful corporations. In political polls, energy conservation, environmental stewardship, and wildlife conservation all score high on the agenda of the voting public; yet the electoral system is tilted in such a way that none of these vital initiatives is a priority. In the meantime, our government continues to spend our grandchildren's money on bank bailouts, subsidies for the most wasteful and polluting industries, and war, war, war.

We find ourselves barreling down a path that is predictable from a philosophy of radical individualism. Everyone labors under the myth that each of us is freely pursuing our own self-interest, but the system is easily gamed, and those who cheat on a grand scale have amassed huge fortunes while masses of people who play by the rules don't have a chance.

A global tragedy of the commons looms large. Humanity may never be able to organize itself to turn aside from the path to ecocide/suicide. But if there is hope for us, it lies in democracy.

ZPG—A Long-Term Future for Humanity Must Be a Future Without Growth

Do humans plan to be around as long as the dinosaurs? Dinosaurs reigned over Earth for two hundred million years. Such is the logic of exponential growth that even if we managed to slow population growth to 0.0001 percent per century—a million times smaller than population growth of the recent past—then long before we reached two hundred million years, the biomass of humanity would exceed the mass of the visible universe. This is but an amusing mathematical demonstration of the obvious fact that long-term survival requires a stable population. ZPG=zero population growth.

In his story "2 B R 0 2 B" (pronounced *2 B R Nought 2 B*), Kurt Vonnegut depicts a world in which aging has been abolished. The title refers to the phone number to reach the Federal Bureau of Termination. To stabilize the population, births are not permitted without deaths. Since there is no aging and disease is rare, deaths occur only by accident, infanticide, and suicide. A painter in the story is two hundred but looks thirty. While working on a mural in a hospital, he witnesses the protagonist, a possible father-to-be of triplets, kill himself, the hospital's chief obstetrician, and a woman posing for the painter. The father's double homicide and suicide occur because he doesn't want to kill his wife's father, a potential grandfather who has volunteered to die to "make room" for just one of his triplets. Killing three allows all three to be born and keeps the population stable. The artist has been painting a social realist mural depicting the metaphorical garden of a population-stabilized society. Descending from a stepladder after the murders, the painter decides against killing himself immediately and instead puts in a call to the story's eponymous number. "Thank you, sir," replies the receptionist at the municipal gas chambers of the Federal Bureau of Termination (also known as "the Catbox" and "Kiss Me Quick"). "Your city thanks you; your country thanks you; your planet thanks you. But the deepest thanks of all is from future generations."

We Call It the Future Because We Don't Know What's Gonna Happen

In the Hopi language and certain others, the future, which we picture as spread out before us, is referred to with words allied to the preposition "behind." This is because we can see the past through our memories, while the future is obscure, unknown, as if it were behind our backs. We cannot see the future, but that won't stop us from trying. The classic mistake is to extrapolate too directly from the past. We don't foresee the upheavals, the disruptions that change the rules of the game. Visionaries like Aubrey de Grey and Ray Kurzweil do their best to err in the opposite direction. But it is one thing to say with confidence, "The future will not be like the past," and another thing to foresee just what those disruptions will be.

The future—that almost magical word is vast, hopeful, fearsome, and contingent on our present actions and attitudes in ways that are maddeningly hard to discern. As our understanding of aging increases, and the stakes concomitantly augment, we would dearly like to know just how longer life spans will play out in a social order that is already changing more rapidly than we can account.

The openness and richness of the past suggest the openness and richness of the future. Eschatology, the discipline and doctrines of what is to come, has traditionally been a religious concern but deserves a broader constituency.

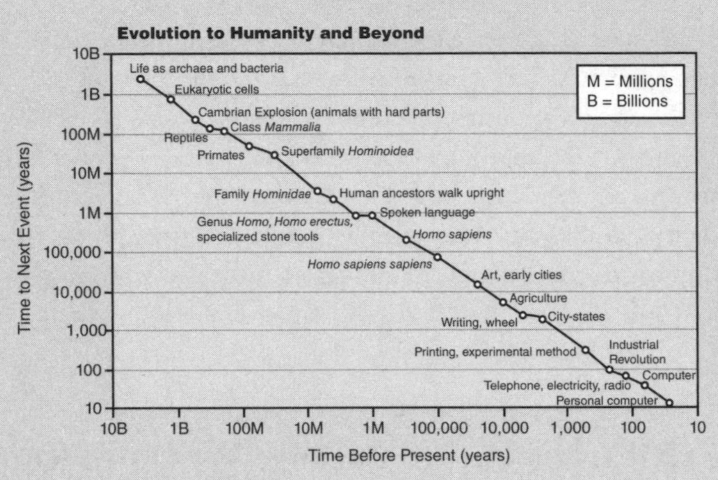

Fig. 4. The history of life, civilization, and information, showing exponentially accelerating change.

The point of the graph is in the logarithmic scale on the bottom. Notice that one billion years on the left occupies the same interval as one million years in the middle of the plot and ten years on the right. Important turning points are arriving faster and faster. Kurzweil claims that the acceleration of history is reaching a cusp in our lifetimes, and he has an idea what that cusp will look like. Human intelligence is destined to merge with machine intelligence over the coming decades, to spread through the universe and dominate it.

The "Singularity" about which he writes is the point at which machines become capable of designing and building new machines, with each generation smarter than the last. A new and very rapid kind of evolution will thus be initiated, with consequences about which we can only dream.

A generation ago, Kurzweil was a pioneer of three artificial intelligence technologies that have matured into consumer products: music synthesis, optical character recognition, and voice recognition. Now, Kurzweil works as director of engineering at Google. He is also a cancer survivor and a prominent advocate of life extension.

We are making exponential progress in every type of information technology. Moreover, virtually all technologies are becoming information technologies. If we combine all these trends, we can reliably predict that, in the not too distant future, we will reach what is known as The Singularity. This is a time when the pace of technological change will be so rapid and its impact so deep that human life will be irreversibly transformed. We will be able to reprogram our biology, and ultimately transcend it. The result will be an intimate merger between ourselves and the technology we are creating . . . Some commentators have questioned whether we would still be human after such dramatic changes. These observers may define the concept of human as being based on our limitations, but I prefer to define us as the species that seeks—and succeeds—in going beyond our limitations. Because our ability to increase our horizons is expanding exponentially rather than linearly, we can anticipate a dramatic century of accelerating change ahead.

—RAY KURZWEIL

A Thousand Years

Back in chapter 1, we offered the actuary's definition of aging: aging is the *increase* from year to year in the probability of dying. Without aging, the probability of dying would never increase, but that doesn't mean it would be zero.

In modern Western societies, the probability of dying in the twentieth year of life is about 1/1,000. A nineteen-year-old thus has a 999/1,000 chance of surviving to celebrate her twentieth birthday. This means that if she never aged, the probability of dying of some disease or a car accident or suicide or some other misadventure would continue to be 1/1,000 for each succeeding year. Her life expectancy would be one thousand years.

All that antiaging medicine can do is to reach for this ideal of one thousand–year life expectancy. Can life expectancy be expanded beyond this? Yes, but advances in biomedicine won't help us get there. We should look instead to safer traffic flow, better workplace safety, social supports to prevent suicide, and full legal implementation of the Kellogg-Briand Pact.*

Some Questions Are Easy

Policy questions regarding the individual benefits and collective risks of antiaging medicine are complex, to be sure. But two closely related questions are as plain as a pikestaff, and we have yet to get them right. Before we wrestle with the difficult trade-offs, let's embrace the win-win policies with gusto.

First, we can live much lighter on the land than we do now. We have only begun to tap the vast potential of recycling, reuse, and energy conservation. Perverse government incentives continue to subsidize energy extraction and a throwaway economy. There is no excuse for them. It's not only more environmentally responsible to conserve energy and materials, it's also cheaper. (If you want to read more about this, we recommend literature from the Rocky Mountain Institute and books by its director, Amory Lovins.)

Second, the world must move toward a culture of smaller families. Encouragement to bring children into the world has long outlived its usefulness, but examples are still rampant. All fundamentalist religions encourage fertility. Western countries, led by the United States, offer tax advantages to parents. It is unforgivable that on a planet with an overextended human family, Germany, Japan, Russia, Italy, and Taiwan all offer cash payouts to encourage parenthood. For many of America's poorest teenage girls, becoming an unwed mother is the only clear path to a subsistence income. These policies can be explained by persistent religious traditions, by the capitalist imperative to grow, and by racism; but they cannot be excused. To encourage childbearing instead of immigration and adoption is a conceit of global irresponsibility. Some of the most effective family planning programs work with indigenous cultures of Africa and elsewhere, using popular radio and TV to change

*If you are not familiar with this slice of history, you might be interested to look it up. Forty years before Simon and Garfunkel, the world really did agree to put an end to war.

attitudes about family size. Along the way, they empower women and improve their status in the community. Population Media Center and Population Communication are small nonprofits that have been doing this work on a shoestring for decades.

A Book, Like a Symphony, Should End Where It Began

This book began with the image of a small boy running into his parents' bed because he was terrified that one day he would cease to be. And now I ask myself, do I love life so much that I can't get enough of it? Or do I fear death so much that I want to put it off, not to have to think about it?

Several years ago, I confided to a friend and colleague in antiaging science that all my life I had experienced a terror of death that I thought carried beyond the ordinary, nearly universal fear. He told me I was not alone, that it was an open secret in the life-extension community that inordinate fear of death was common.

Freud came of age with the first post-Darwinian generation, and of course he was no stranger to evolutionary ideas. Much of Freud's understanding of unconscious human motivation fits comfortably with the idea of a brain evolved for propagation and self-preservation.

But late in life, Freud's research took a curious turn as he discovered, described, and then documented a class of motives that he called *Todestrieb*, "death drives." While the thrust (so to speak) of his early work was easily reconcilable with a Darwinian view of psychology, the pull of Thanatos was something stranger. He described it first as a compulsion among his patients to repeat old traumas, to re-create situations that would remind them of some terrifying event in their past. Only after many years did he generalize this with the thesis that "from dust we came," and the inanimate within us continues to exert a force that pulls us to return to the earth and to a lifeless state.

Scholars looking back on the development of Freud's idea note that the examples that he cited hardly justify the hypothesis of a generalized psychological pull toward death. But perhaps Freud's own life tells us where the idea came from. Freud continued to smoke a full box of cigars every day for years after his jaw was surgically removed because of mouth cancer. When the pain of that cancer became more than he could bear, he died via physician-assisted suicide with a morphine overdose at age eighty-three. I like to speculate that in expounding details of the Death Drive Theory, Freud was really writing from his own introspection, that he was fatigued after a long career, fighting one battle after another to achieve academic acceptance of his research. He lived to see Nazis burning his books, and he fled Europe in the final year of his life. Perhaps he felt ready for the long sleep.

Today, we are saddened but not surprised when, toward the end of a long and full life, an elder expresses a readiness to let go of life. Could this be a manifestation of a "Darwinian Death Drive," ancestor to Freud's *Todestrieb*?

So is that why I am interested in aging and death? Could it be that fear is the only reason?

I really don't know. But the truth is that since I began, in 1996, the study of life extension and implementation of my findings in my own life choices, I have received two blessings. The lesser blessing is that I am healthier, more active, more alive, more energetic, and more flexible, and I have more endurance than people my age are wont to expect from their bodies. This boon has come through some combination of my genes, my community, and my personal efforts, and I am loath to quantify proportions to the three contributors. The greater blessing has been a lifting of the pall of fear. The terror that had always been just over my left shoulder, threatening to seize my consciousness and disable me at any time, is no longer stalking me. I look for the goblin and discover that he is now a shadow in my memory—perhaps a return visitor in my future, but for now, he seems far away.

Stephen Cave (in a book and a TEDx talk) has offered us some clear insights in an area where clarity has been elusive. He reminds us that willful self-deception is part of our nature, and nowhere is it more active than in denial of death. Historically, humans have sought relief in four types of stories:

1. An elixir of youth, that can provide immortality for this body
2. Life after death, reincarnation
3. Immortality of a soul, independent of the body
4. A lasting legacy in our ideas and good works

In the past, people looked to magic or to alchemy or to all varieties of religious tradition for a release from the fate we did not wish to accept. You and I are heirs to an Enlightenment tradition, and so we look to science for that same release. The need to delude ourselves is no less strong, and science does a more convincing job for us than any of these belief systems from the past.

The thing that I find curious, from the perspective of my own psychology, is that my pursuit of longevity through science has been so unreasonably effective at quieting my fears. I believe that I have bought an extra decade of life through my efforts, and I have reason to hope that emerging life-extension technologies will avail themselves quickly enough that I may receive the blessing of another decade or even several decades from advances in science. The paradox is that, by any measure, I am closer to death now than when my three-year-old self ran shaking into his mother's bed in the middle of the night; and yet terror of the Void is so much further away.

In the fourth century BC, Epicurus* bequeathed us the thought that fear of death was irrational. "So long as I am, death is not; and when there is death, there will not be me." Wittgenstein offered a twentieth-century formulation of the same thought:

* τὸ φρικωδέστατον οὖν τῶν κακῶν ὁ θάνατος οὐθὲν πρὸς ἡμᾶς, ἐπειδήπερ ὅταν μὲν ἡμεῖς ὦμεν, ὁ θάνατος οὐ πάρεστιν, ὅταν δὲ ὁ θάνατος παρῇ, τόθ́ ἡμεῖς οὐκ ἐσμέν.

Death is not an event in life: we do not live to experience death. If we take eternity to mean not infinite temporal duration but timelessness, then eternal life belongs to those who live in the present. Our life has no end in just the way in which our visual field has no limits.

But I didn't find this idea comforting when I was three, and I am not comforted by it today. More helpful for me is evolutionary understanding of the fear of death, where it comes from, how it asserts its power over me. I think that (in this case) fear of death is well explained by selfish gene theory. A powerful aversion to death has been implanted in our metabolism as a way to help preserve the very genes that are the basis of that aversion and to increase the likelihood those genes will be transmitted to the next generation. The terror comes from a rush of adrenaline, and the sense of urgency is evolved to help us in times of mortal danger to have extra reserves of strength and to assure that we forget everything else but the fight for our lives.

The hormonal basis for fear of death evolved in animals that were not yet sufficiently developed to have an understanding of the long-term future. They lacked awareness of their eventual mortality, and thus the terror appeared only in brief episodes, when it provided (we may imagine) energy and motivation that were useful for escaping immediate danger. But in modern humans, that primal terror has diffused through our awareness, and it can be a continual, low-level presence, a demon who rides on our left shoulder. As Cave says, "This is our curse. It's the price we pay for being so damn clever." As a chronic condition, terror is not adaptive—far from it. Living with an awareness of our mortality may be a way to motivate some of our noblest efforts, but living with terror can only be paralysis. The hormones and the sensations they bring on are a vestige, a response that was adaptive in the jungle but not in the context of our present culture.

There is no escaping death, but perhaps we can escape from fear.

Epilogue

Any scientist can tell you that the only legitimate function of experiment is to *dis*prove a theory. The best theories that we have are the ones that have not been disproven yet. We feel confident that the evidence we have presented here disproves some popular ideas about aging and evolution. We have argued:

- that the body ***does not*** have to wear out,
- that aging ***has not*** evolved as a side effect of fertility genes, and
- that the selfish gene ***cannot*** explain aging or sex or stable ecosystems.

Does that mean that our theory is correct, that aging exists in order to stabilize ecosystems, to level the death rate in good times and hard times? We are much less certain of this, but we think it is a hypothesis worth exploring.

Both economic theory, as Adam Smith's "invisible hand," and neo-Darwinian theory, as "survival of the fittest," depend on individuals pursuing their own selfish interest with no regard for others. These great paradigms are one-sided. They neglect the brilliant success of alliances that have evolved to work together as concerted units. From primordial bacteria trading genes and merging bodies to protists evolving into the ancestors of plants and animals, to fungi spreading beneath the ground to nourish and recycle trees, to groups of neo-Darwinists buttressing one another on the comment sections of social media sites—natural selection rewards both competition *and* cooperation. When one side of this dialectic is emphasized at the expense of the other, there can be practical consequences as negative for society and human potential as the theoretical consequences are for understanding ourselves in terms of evolutionary theory.

A central theme of this book is that excessive growth leads to extinction, and natural selection has built safeguards for ecological stability into the genome of almost all animal species. The irony is that while the human race may not have the commitment or the democratic structures that would support intelligent management of our population growth, nature is an old hand at the process, having long since mastered the recurring problem of rampant growth. If our human tools—intelligence and large-scale cooperation—cannot do the job, she will resolve the dilemma of human growth in her own way.

We begin to recognize that our individual lives do not belong to us alone but to our species—and to the other species upon which human life interdepends and to life itself. There is a social contract that predates human community by several billion years, which offers life to each individual as a loan, and not a gift.

We have seen that aging is not accidental or fated but has evolved as part of nature's four-billion-year construction kit. That kit continues to evolve, and if we play our cards right, our brand of living organization may continue along for the wild ride to come. We are succeeding at delaying debilitating disease, forestalling bodily decay, extending the prime of life. We have a reasonable expectation that this progress will continue, perhaps at an accelerating pace. But we have not addressed, and science can never address, the fundamental mortality of our physical bodies. Infinity is not part of physics, let alone biology. And humans have sought consolation for our mortality since such time whereof the memory of man runneth not to the contrary. Reaching beyond science, we turn in this pursuit to the persistence of our good works, to family legacy, and also to creations, meditations, superstitions, politics, religion, self-deception, and high art, individually and collectively, each after our own fashion.

ACKNOWLEDGMENTS

DS would like to thank the many people who provided feedback or support while working on this book, including Andre Khalil, Mike Carragher, Tori Alexander, Susan Black, Angus Fletcher, Lester Grinspoon, David Lenson, Olav Bryant and Tara Grover Smith, Paul, Arthur, and Elaine Mange, John and Deb Lapaire, Diane Alexander, Nick Seamon, Brian Viveiros, Claire Brault, Lois Brynes, Paul Levy Bryant, Steve Shaviro, Paul Chefurka, Georges Borchardt, Ricardo Guerrero, James MacAllister, Kurt Johnson and Stephen H. Blackwell, Jane Shevtsov, Charles Hall, John Grady, Jim Brown, Mitch Mignano, Francesca Ferrando, Cristina Iuli, Emily Wadham, Lynn Margulis, Tonio Sagan, Meghan Murphy, Melody Meoskowicz, Joanna Bybee, Bruce Clarke, Kevin LeGrandeur, Natasha Vita Moore, Harold Channer, Rachel Lorraine Young, Ryan Kronewitter, Stan Maron, Corey Reed Smith, Natasha Myers, Natalie Loveless, Eben Kirksey, Astrid Schrader, Elizabeth Johnson, Erik Middleton, Steve Wilson, Jon Lomberg, Jennifer Margulis, Debbora Battaglia, Donna Haraway, William I. Thompson, Nora Bateson, Kee Dewdney, Lloyd Demetrius, Tom Kirkwood, Walter M. Bortz II, Leonard Hayflick, and Inocencio Higuera and colleagues at CIATEJ in Guadalajara.

This book was built on a great body of thought and diligent observation that went before, and benefited crucially from feedback all along the path of its making.

JJM would like to thank Harriet Mitteldorf, John Pepper, Mark Bernstein, Enid Kassner, Court Daspit, Gustavo Barja, Aubrey de Grey, Bill Andrews, Jeff Bowles, Meng-Qiu Dong, Charles Goodnight, David Wilson, Cynthia Kenyon, Michael Fossel, George Martin, Daniel Promislow, Justin Travis, Devereux Chatillon, Vladimir Skulachev, João Pedro de Magalhães, Bruce Ames, Rhonda Patrick, Giacinto Libertini, and readers of his blog who have caught his mistakes before they could travel too far, and have pointed him to research of which he was unaware.

Dorion and Josh would both like to give a special thanks to our agent, Gillian MacKenzie, as well as Allison Devereux and the rest of her excellent agency, as well as to the great team at Flatiron Books, including Colin Dickerman and Jasmine Faustino, and the copy editors, Sara and Chris Ensey, whose professional expertise and attention to detail have contributed so much to making this book what it is.

GLOSSARY

Aging The biological definition of aging can be any change in the body that comes with age, including growth, development, and puberty. Some biologists reserve the word **senescence**, q.v., to refer to just those changes that involve loss of function and breakdown later in life, but modern usage is that **aging** is synonymous with **senescence**. Demographers have their own definition: aging is the increase in probability of death that comes with changes in the body. This allows for the possibility of **negative senescence**, q.v., or aging backward, if the probability of dying goes down with age.

Antagonistic Pleiotropy (AP) Most genes act in more than one way. One gene can have several effects. This is called *pleiotropy*. When some of these effects are beneficial and other effects of the same gene are detrimental, the gene is fighting with itself. This is *antagonistic pleiotropy*. In this case, natural selection has to either forgo the benefit or else make a bargain with the devil. If the benefits come early in life and the costs arrive late in life, that sweetens the deal with the devil. For example, the pleiotropic gene may make many copies of itself by enhancing fertility early in life. By the time it kills the individual that is carrying the gene, the gene has already passed itself to the next generation in multiple copies. According to a prominent and popular theory, this is the way in which aging has evolved.

Antioxidants **Free radical scavengers**. These are molecules that quench or neutralize **free radicals**, q.v., by giving their electrons something to attach to. Examples of antioxidants include glutathione (GSH), superoxide dismutase (SOD), ubiquinone (CoQ10), polyphenols (found in tea), and vitamin E.

Apoptosis **Programmed cell death**. The "purposeful" and orderly destruction of a cell. It is triggered by the mitochondrially orchestrated deployment of free radicals. Apoptosis is implicated in **aging**, q.v., via its role in **sarcopenia** (muscle loss), Parkinson's disease (dementia), and **thymic involution**, q.v. (a decline in immune ability traced to shrinking of the **thymus**), as well as *atresia*, which destroys a woman's eggs through the duration of her fertility.

Black Queen The theory that aging evolved to increase population diversity and the adaptability of populations. The name was coined by one of us (DS) after reading an academic article written by the other (JJM) about Red Queen theory. The Red Queen is a mainstream theory for the evolutionary origin of sex, based on popular diversity and adaptability. In this volume, we argue that these factors played a central role in the evolution both of sex and of aging, but that (quite probably) neither could have prevailed in an uphill battle against individual selection, were it not for another broad group benefit, which we call the **Demographic Theory of Aging**, q.v.

Caloric Restriction (CR) The surprising finding, known since the 1930s, that eating less, even to the point of near-starvation, can increase longevity and improve health. CR is an example of **hormesis**, q.v., and fits well with this book's theme of **demographic homeostasis**, q.v.

Declining Force of Natural Selection ("Medawar Hypothesis") Natural selection "cares more" about young individuals than old individuals. What happens to an individual later in life has less effect on the prospects of its genes than what happens earlier. This is true for two reasons: First, the individual might die for external reasons that have nothing to do with its age, and so it never reaches old age. Second, after an individual reproduces, its offspring carry its genetic legacy into the third and fourth generations, and this can be more important than what happens to the original individual. British biologist Peter Medawar first put forward the idea of the Declining Force as an evolutionary explanation for aging. Perhaps so few animals reach old age in nature that aging is actually "invisible" to natural selection? Later, field studies showed that this is not the case, so Medawar's observation is true, but it cannot be the whole story.

Demographic Homeostasis (DH) A stable, sustainable configuration of population that can restore itself when population rises too high—and recover when population falls too low—without the violent population swings that can lead to extinction. The idea that natural selection can create demographic homeostasis in a Darwinian process was an implicit theme of a controversial book by V. C. Wynne-Edwards. The mainstream of neo-Darwinian theory rejected this idea in the 1960s and 1970s, even after Michael Gilpin effectively built a theoretical foundation under Wynne-Edwards's observations. A central theme of this book is that aging evolved based on its contribution to DH.

Dioecious (also diecious) An adjective describing species that have separate male and female forms. Distinguished from organisms (bacteria, rotifers) that reproduce by cloning, and others (worms, snails, flowers) that have both male and female organs in the same individual.

Epigenetics The genome (DNA) is the book of information for everything that a living being needs to do. The same DNA inhabits every cell of the body and does not change from womb to tomb. But different cells read selectively from the book of DNA in order to accomplish their individual tasks. Parts of the genome are active and parts are hidden, and the portion that is active varies from one organ of the body to another and also changes with the time of life. Epigenetics is the control of which genes are active where and when. In terms of biochemistry, this is accomplished through tightly spooling some portions of the DNA, hiding them from sight, while opening other portions. Chemical decorations on the DNA control this process—for example, methyl groups and acetyl groups. A great deal of the body's signaling is devoted to programming the genome with these controls.

Epistasis The context-dependence of the effect of a gene. Almost all genes depend for their action on other genes. When the term was first invented, it was not yet known that DNA was the biochemical basis for genetics, and the science was developed in the abstract. Later, it became apparent that epistasis is not at all exceptional, and that genes quite generally are organized in networks and hierarchies. In fact, there are only a very few genes that are *not* epistatic, for example, the gene that causes sickle cell anemia.

Eustress The surprising positive effect of certain kinds of stress on health and life span. Akin to **hormesis**, q.v.

Exponential Growth Growth that follows an exponential curve, with a characteristic doubling time. In other words, it takes the same amount of time to grow from one to two as from ten to twenty or one million to two million. Mathematically, any time the rate of increase of a quantity is proportional to the quantity itself, you have exponential growth. Exponential growth is famous for being a sleeper—it may look for a long while as though nothing is happening, and then suddenly it shoots up like a rocket.

Extinction (or Wipeout) Refers to the disappearance of species, populations, or ecosystems—that is, of groups. Crucial to the operation of **group selection**, q.v.

Fitness For Darwin, the attributes, varying according to climate and geography and ecological context, which make one organism more successful than another in evolutionary competition. As evolution moved from being a descriptive to a mathematical science in the twentieth century, fitness was recast with a rigid definition in terms of reproduction that led to the selfish gene version of evolutionary theory, or **neo-Darwinism**, q.v. Many modern evolutionists, including the authors of this book, believe that this is an oversimple account of how natural selection really works.

Free Radicals Also known as ROS (reactive oxygen species), or simply radicals, free radicals are atoms or molecules with at least one unpaired electron that are quick to react to form a more stable molecule. They can be neutralized by free radical scavengers (a.k.a. **antioxidants**, q.v.) or they can cause damage. Deployed by the immune system, free radicals can help the body—for example, by destroying pathogens or damaged tissue. A popular theory of the late twentieth century held that free radicals might be a cause of aging, and that antioxidants might be a panacea. We now know that antioxidants generally don't increase life span, and sometimes can actually be detrimental. This is because free radicals are also potent signal molecules that invite the body to repair itself.

Free Radical Scavengers **Antioxidants**, q.v.

Free Radical Theory of Aging See *Mitochondrial Free Radical Theory of Aging.*

Genetic Load Natural selection is constantly doing its work, amplifying the good mutations, weeding out the bad. Each mutation is an experiment, and most experiments by far turn out to be dead ends, but that's the price to be paid for an ongoing search that, every once in a while, hits pay dirt. At any given time, there are a lot of bad mutations that have arrived recently in the genome and have not yet been weeded out. And by the time they *are* weeded out, there will be new detrimental mutations. Fitness of every individual is depressed a bit by carrying a load of bad mutations that have not yet been weeded out.

Group Selection (GS) When theorists fifty years after Dawin first cast evolution as a quantitative theory, they used the simplest possible assumptions to keep the math tractable. The theory was based on one gene at a time, competing in a population of fixed size. Over the decades of the twentieth century, the theory of the selfish gene became a core of neo-Darwinism, and widely taught for its mathematical elegance. By 1960, a whole generation of evolutionary scientists had learned that "this is the way evolution works." This was the era in which the evolution of cooperation and symbiosis first became subjects of study.

In fact, cooperation evolves because groups and communities and entire ecosystems compete among themselves, just as individuals do. Groups, like individuals, are subject to Darwinian selection. But because of the way people had been trained, the idea of competing groups seemed alien to them. They did not remember that "one gene at a time" and "constant population size" were no more than simplifying assumptions introduced when all theoretical calculations had to be performed with paper and pencil. The idea that natural selection operates on many levels at once is called *multilevel selection*, q.v., or MLS. MLS should be the most natural assumption in the field, but, by this historical accident, mainstream theory had been developed based on groups that remained fixed and stable. Thus a special skepticism grew up around "group selection" that continues to prejudice theory in the field to this day.

Hayflick Limit Human cells can be grown in lab cultures for forty or fifty generations before they languish and die. This is known as the Hayflick limit. Decades after Hayflick's original demonstration of this fact, the biochemical basis of the Hayflick limit was discovered to be the shortening of telomeres, q.v. Cancer cells can go on multiplying indefinitely, and so can cells of plants and some animals. These are cells that are programmed to maintain telomere length with telomerase each time they divide.

Hormesis The metabolism responds to some stressors and challenges with an overreaction. Sometimes an animal or plant can be healthier and live longer with stress than without. Examples of hormesis include hunger (see **caloric restriction**, q.v.), exertion, heat, cold, some toxins, and most doses of radiation.

Inflammation The body has natural defenses that destroy diseased tissue or invading cells with highly reactive molecules, **ROS**, q.v. Early in life, inflammation serves exclusively a protective function, but later in life, inflammation is co-opted to destroy the body's own healthy tissue. This is one of the primary mechanisms of programmed aging.

Iteroparity Refers to animals or plants that reproduce episodically, more than once in a lifetime. Distinguished from **semelparity**, q.v., in which each individual reproduces once and then (almost always) dies.

Metformin Glucophage, a diabetes drug that incidentally reduces risk of cancer and heart disease. Mice that are fed metformin live longer, and it has been proposed that metformin may be a modestly effective antiaging drug.

Mitochondrial Free Radical Theory of Aging (MFRTA) Mitochondria, the cell's energy factories, produce copious **free radicals**, q.v., **ROS** that can damage the mitochondria themselves as well as the cell's other delicate machinery. According to MFRTA, this is a primary cause of aging. In fact, this damage seems to be an effect of aging rather than a cause, and the situation is complicated by the fact that these same **ROS** act as signal molecules that can prompt the cell to repair itself better than new.

Mutation Accumulation (MA) This is a theory of aging based on **genetic load**, q.v. The idea is that some mutations cause harm in an individual, but only late in life. If the harm occurs after the individual has had a chance to reproduce, then the genes may persist for a long while before natural selection eventually weeds them out. According to the MA Theory, it is these genes that are the reason for aging generally.

The MA Theory lost credibility when it was discovered that many animals in nature do die of old age, so aging has a large effect on fitness. The theory lost what remained of its credibility when the ancient genetic basis of aging was discovered.

Natural Selection In every population, some individuals are better able to survive and to reproduce than others. Their differential success is called "natural selection," and is the basis of Darwinian evolution. Of course, some groups and communities also fare better than others in natural competition, and this is the controversial subject of **group selection**, q.v.

Negative Senescence **Reverse aging**, q.v. Normal senescence is the loss of fertility with age, along with a frailty that leads to increased risk of death. But some animals and many trees grow larger and more fertile with each passing year, and this is called "negative senescence."

Neo-Darwinism A mathematical version of Darwin's theory of evolution by **natural selection**, based on the simplifying assumptions that populations are fixed in size and genes are (mostly) independent agents. After its origins in the early twentieth century, this version of the theory grew to be the mainstream of evolutionary science, based on its mathematical elegance and its success in accounting for laboratory experiments in evolution.

NSAIDs Nonsteroidal anti-inflammatory drugs. Examples include aspirin, ibuprofen, naproxen (Aleve), and some stronger arthritis drugs.

Pleiotropy Pertaining to a situation where a single gene controls two or more unrelated traits, sometimes staggered over time. When the term was first invented, it was not yet known that DNA was the biochemical basis for genetics, and the science was developed in the abstract. Later, it became apparent that pleiotropy is not at all special, and that almost all genes have multiple effects. **Antagonistic pleiotropy**, q.v., is the case where a single gene confers fitness in some respects but has detrimental side effects.

Population Homeostasis See *Demographic Homeostasis.*

Programmed Cell Death **Apoptosis**, q.v. See also *Mitochondrial Free Radical Theory of Aging.*

Protists Protists are single-celled eukaryotes. They are much larger and more highly-structured than bacteria, which are prokaryotes. Almost all contain cell nuclei (with chromosomes) and mitochondria. Examples of protists include amoebas, paramecia, and other ciliates, and the malarial parasite plasmodium. Although crucial in evolution—aging and our kind of sex evolved in them—protists have historically received little attention except when they've caused diseases.

Red Queen In mainstream evolutionary theory, sexual reproduction is a conundrum. Sex is bad for the selfish gene. There is no full resolution, but one of the most prominent theories holds that large, multicelled organisms must constantly change their genomes, lest they be too easy a target for smaller, fast-reproducing parasites (viruses and bacteria). The name "Red Queen" comes from a line of Lewis Carroll's about running as fast as you can just to stay in one place. (The name of our own theory, the "Black Queen" is a riff on this theme.)

Reverse Aging Some hydrozoans and beetles under stress actually revert to earlier life stages, instar or larval forms. F. Scott Fitzgerald's story of Benjamin Button imagines a human example. See also *Negative Senescence*.

ROS (Reactive Oxygen Species) a.k.a. **Free radicals**, q.v.

Second Law of Thermodynamics The physical principle, formulated in the nineteenth century, that heat spreads out, that rough edges are worn smooth, and that fast-moving objects are slowed by friction. A quantity called entropy was defined by Rudolf Clausius in the context of heat engines and limits to their efficiency. Entropy measures the extent to which energy is dissipated, and the entropy always increases in any isolated system (not interacting with its environment).

Living systems are uniquely adapted to skirt the Second Law, but not to violate it. All living things take in free energy from the environment and dump their entropy out as waste. As a result, there is no necessity for the entropy of an individual plant or animal to increase (otherwise growth and development would be impossible).

Misapplication of the Second Law to aging has been an enduring misconception in the field of aging science.

Semelparity Refers to animals or plants that reproduce once and then (almost always) die. Distinguished from **iteroparity**, q.v., in which each individual reproduces episodically, more than once in a lifetime.

Senescence This is the biological term for what we commonly call "aging." Senescence can mean either a loss of function with age, especially fertility, or an increase in risk of mortality.

Superoxide Dismutase (SOD) An enzyme that defangs the highly destructive superoxide radical (chemical formula O_2^-), turning it into less toxic hydrogen peroxide or to ordinary oxygen (O_2).

Thymic Involution The thymus is a small gland at the top of the breast bone that trains immune cells in the blood to do their job (white T cells, "T" for "thymus"). The thymus is large and active in youth, and it atrophies ("involutes") with age with consequences that are increasingly devastating for immune function.

Tragedy of the Commons It sometimes happens that individuals in a community are each making rational choices, given their circumstances, but that the collective result of their behaviors is a disaster. Of course, the devastation visits every individual, and the fact that they each behaved rationally offers no comfort. The phrase comes from a parable published in *Science* magazine in 1968 by Garrett Hardin. This is the opposite of Adam Smith's "invisible hand" that brings about a global optimum that arises magically from behaviors that are individually selfish.

Universal Wear and Tear Nonliving things are subject to running down over time, an increase in entropy according to the **Second Law of Thermodynamics**. Living things are not subject to this law because they can extract free energy from the environment. But the erroneous application of this idea to living things, and to aging in particular, has sown a great deal of confusion.

Weismann Theory August Weismann (1834–1914) was the first significant evolutionary theorist after Darwin. His ideas about aging are frequently caricatured as "the old must die to make room for the young." What he actually wrote was more difficult to pin down, but his emphasis was on removal from the population of individuals that become damaged through a lifetime of hazards and accidents.

NOTES

Preface: What This Book Is About

6 *In addition to AGE-1* Johnson, T.E., *Increased life-span of age-1 mutants in Caenorhabditis elegans and lower Gompertz rate of aging.* Science, 1990. 249(4971): p. 908–912.

6 *Worms with the AGE-1 mutation* Johnson, T.E., P.M. Tedesco, and G.J. Lithgow, *Comparing mutants, selective breeding, and transgenics in the dissection of aging processes of Caenorhabditis elegans.* Genetica, 1993. 91(1–3): p. 65–77.

6 *There are now hundreds* Stearns, S.C., *The Evolution of Life Histories.* 1992, Oxford; New York: Oxford University Press. xii, p. 249.

7 *"killer" genes* Guarente, L., and C. Kenyon, *Genetic pathways that regulate ageing in model organisms.* Nature, 2000. 408(6809): p. 255–262.

9 *Competition . . . must be regulated* Wilson, D.S., *The new fable of the bees: multilevel selection, adaptive societies, and the concept of self interest.* Evolutionary Psychology and Economic Theory, Advances in Austrian Economics, ed. Roger Koppl, vol. 7, 2004, Amsterdam, Elsevier Ltd., p. 201–220.

11 *the ways in which a community functions as a collective* Margulis, L., *Origin of evolutionary novelty by symbiogenesis.* Biological Evolution: Facts and Theories: A Critical Appraisal 150 Years After "The Origin of Species." Rome: Gregorian and Biblical Press, 2011. 312: p. 107–114.

11 *evolutionary science today is missing something* Woese, C.R., *A new biology for a new century.* Microbiology and Molecular Biology Reviews, 2004. 68(2): p. 173–186.

Prologue: Your Inner Stalker

15 *"intellectual hero"* Hamilton, W.D., *Narrow Roads of Gene Land: Volume 2: Evolution of Sex.* 2001, Oxford: Oxford University Press.

16 *our ability to easily make organisms live longer* Pletcher, S.D., *The modulation of lifespan by perceptual systems.* Annals of the New York Academy of Sciences, 2009. 1170(1): p. 693–697.

16 *insulated from the mere* smell *of food* Libert, S., et al., *Regulation of Drosophila life span by olfaction and food-derived odors.* Science, 2007. 315(5815): p. 1133–1137.

19 *Each of us walks around* Dawkins, R., *The Extended Phenotype: The Long Reach of the Gene*. 1999, Oxford: Oxford University Press.

21 *In "Dr. Heidegger's Experiment"* Hawthorne, N., *Dr. Heidegger's Experiment*. 1897, Doubleday, McClure & Company.

Introduction: How a Lifelong Obsession with Aging and Health Became My Career

26 *A modest load of toxins in the diet* Ames, B., *Dietary carcinogens and anticarcinogens: Oxygen radicals and degenerative diseases*. Science, 1983. 221(4617): p. 1256–1264.

26 *we're likely to live longer with the toxins* Ames, B.N., R. Magaw, and L.S. Gold, *Ranking possible carcinogenic hazards*. Science, 1987. 236(4799): p. 271–280.

27 *animals that lived longer the less they were fed* Weindruch, R., *Caloric restriction and aging*. Scientific American, 1996. 274(1): p. 46–52.

32 *I looked up a paper* Maynard Smith, J., *Group selection*. The Quarterly Review of Biology, 1976. 51: p. 277–283.

33 *I came across a feature* Berreby, D., *Enthralling or exasperating: select one*, in *New York Times*. 1996.

33 *we collaborated closely together* Mitteldorf, J., and D.S. Wilson, *Population viscosity and the evolution of altruism*. Journal of Theoretical Biology, 2000. 204(4): p. 481–496.

1. You Are Not a Car: Your Body Does Not "Wear Out"

39 *A well-accepted theory of aging* Kirkwood, T., *Evolution of aging*. Nature, 1977. 270: p. 301–304.

44 *living things are* open systems Schneider, E.D., and D. Sagan, *Into the Cool: Energy Flow, Thermodynamics, and Life*, 2006. Chicago: University of Chicago Press.

47 *The Orgel Hypothesis* Orgel, L.E., *The maintenance of the accuracy of protein synthesis and its relevance to ageing*. Proceedings of the National Academy of Sciences, 1963. 49: p. 517–521.

47 *there was no appreciable accumulation* Harley, C.B., et al., *Protein synthetic errors do not increase during aging of cultured human fibroblasts*. Proceedings of the National Academy of Sciences, 1980. 77(4): p. 1885–1889.

48 *Denham Harman studied the effects of radiation on mice* Harman, D., *Aging: a theory based on free radical and radiation chemistry*. The Journals of Gerontology, 1956. 11(3): p. 298–300.

50 *antioxidants did not seem to protect the cells or make lab animals live longer* De Grey, A.D., *The Mitochondrial Free Radical Theory of Aging*. 1999, Austin, TX: Springer/Landes, p. 212.

51 *antioxidants were killing their subjects* ATBC, *The effect of vitamin E and beta carotene on the incidence of lung cancer and other cancers in male smokers. The Alpha-Tocopherol, Beta Carotene Cancer Prevention Study Group*. The New England Journal of Medicine, 1994. 330(15): p. 1029–1035.

2. The Way of *Some* Flesh: The Varieties of Aging Experience

61 *evolution has concerned itself more attentively with things that kill us when we are young* Hamilton, W.D., *The moulding of senescence by natural selection.* Journal of Theoretical Biology, 1966. 12(1): p. 12–45.

62 *In a provocative 2004 article, they offer a general "sproof"* Vaupel, J.W., et al., *The case for negative senescence.* Theoretical Population Biology, 2004. 65(4): p. 339–351.

66 *the breadth of nature's ingenuity* Jones, O.R., et al., *Diversity of ageing across the tree of life.* Nature, 2014. 505(7482): p. 169–173.

73 *Rachel Sussman published* Sussman, R., C. Zimmer, and H.U. Obrist, *The Oldest Living Things in the World.* 2014, Chicago: University of Chicago Press.

75 *when he started to feed the worms again, they grew back* Stoppenbrink, F., *Der Einflul herabgesetzter Ern/ihrung auf den histologischen Bau der Siilwasser-tricladen.* Zeitschrift für Wissenschaftliche Zoologie, 1905. 79: p. 496–574.

75 *"the immortal jellyfish"* Bavestrello, G., C. Suommer, and M. Sara, *Bi-directional conversion in Turritopsis nutricula.* Sci. Mar., 1992. 56(2–3): p. 137–140.

76 *If they were deprived of food for many days* Beck, S., *Growth and retrogression in larvae of Trogoderma glabrum 1. Characteristics under feeding and starvation.* Annals of the Entomological Society of America, 1971. 64: p. 149–155.

76 *they don't seem to die on their own* Martinez, D.E., *Mortality patterns suggest lack of senescence in hydra.* Experimental Gerontology, 1998. 33(3): p. 217–225.

79 *Charles Goodnight and I had an idea* Mitteldorf, J., and C. Goodnight, *Post-reproductive life span and demographic stability.* Oikos Journal, 2012. 121(9): p. 1370–1378.

3. Darwin in a Straitjacket: Tracing Modern Evolutionary Theory

88 *sex represents nature's solution of how to promote cooperation* Peck, J.R., *Sex causes altruism. Altruism causes sex. Maybe.* Proceedings of the Royal Society of London, Series B: Biological Sciences, 2004. 271(1543): p. 993–1000.

88 *Sexual sharing of genes* Bell, G., *The Masterpiece of Nature: The Evolution and Genetics of Sexuality*, 1982. Berkeley: University of California Press, p. 635.

91 *in arid, hot regions around the world* Dawkins, R., *The Selfish Gene.* 1976, Oxford: Oxford University Press.

95 *The title of Dobzhansky's essay* Dobzhansky, T., *Nothing in biology makes sense except in the light of evolution.* 1973.

97 *natural population control* Wynne-Edwards, V., *Animal Dispersion in Relation to Social Behaviour.* 1962, Edinburgh: Oliver & Boyd.

99 *a young mathematical biologist* Williams, G., *Adaptation and Natural Selection.* 1966, Princeton: Princeton University Press.

99 *Naturalists were in the habit* Darwin, C., *On the Origin of Species by Means of Natural Selection, or the Preservation of Favoured Races in the Struggle for Life.* 1859, London: John Murray.

100 *The whole Darwinist teaching of the struggle for existence* Engels to Pyotr Lavrov in London. Written: Nov. 12–17, 1875; Transcription/Markup: Brian Baggins; Online Version: Marx/Engels Internet Archive (marxists.org) 2000. https://www.marxists.org/archive/marx/works/1875/letters/75_11_17-ab.htm.

101 *Fisher's magnum opus* Fisher, R.A., *The Genetical Theory of Natural Selection.* 1930, Oxford: Clarendon Press. xiv, p. 272.

102 *When amateur surrealist* Morris, D., *The Naked Ape.* Life, 1967. 63(25): p. 94–108.

4. Theories of Aging and Aging of Theories

108 *What Weismann wrote was* Weismann, A., et al., *Essays Upon Heredity and Kindred Biological Problems.* 2d ed. 1891, Oxford: Clarendon Press. 2 v.

110 *the foundation of the three modern theories* Medawar, P.B., *An Unsolved Problem of Biology.* 1952, London: Published for the college by H. K. Lewis. p. 24.

111 *a later article* Edney, E.B. and R.W. Gill, *Evolution of senescence and specific longevity.* Nature, 1968. 220(5164): p. 281–282.

113 *about 28 percent of deaths* Bonduriansky, R., and C.E. Brassil, *Senescence: rapid and costly ageing in wild male flies.* Nature, 2002. 420(6914): p. 377.

115 *The phrase was introduced by Michael Rose* Rose, M., *Laboratory evolution of postponed senescence in Drosophila melanogaster.* Evolution, 1984. 38(5): p. 1004–1010.

115 *the idea traces to George Williams* Williams, G., *Pleiotropy, natural selection, and the evolution of senescence.* Evolution, 1957. 11: p. 398–411.

116 *Senescence should always be* Ibid.

117 *such a small number of* Ibid.

123 *The scientific significance* Rose, M.R., *Laboratory evolution of postponed senescence in Drosophilia melanogaster.* Evolution, 1984. 35(5): p. 1008–09

124 *The flies that Rose has bred* Leroi, A., A.K. Chippindale, and M.R. Rose, *Long-term evolution of a genetic life-history trade-off in Drosophila: The role of genotype-by-environment interaction.* Evolution, 1994. 48: p. 1244–1257.

125 *to test the MA Theory* Promislow, D.E., et al., *Age-specific patterns of genetic variance in Drosophila melanogaster. I. Mortality.* Genetics, 1996. 143(2): p. 839–848.

125 *correlations between fertility and longevity, to test for AP* Tatar, M., et al., *Age-specific patterns of genetic variance in Drosophila melanogaster. II. Fecundity and its genetic covariance with age-specific mortality.* Genetics, 1996. 143(2): p. 849–858.

128 *paper proposing the Disposable Soma Theory* Kirkwood, T., *Evolution of aging: how genetic factors affect the end of life.* Nature, 1977. 270: p. 301–304.

129 *the "sproof" article* Vaupel, J.W., et al., *The case for negative senescence.* Theoretical Population Biology, 2004. 65(4): p. 339–351.

131 *Finch interprets this to mean that* Finch, C.E., *Longevity, Senescence and the Genome.* 1990, Chicago: University of Chicago Press.

131 *His 2007 study of seventeen mammal* Ricklefs, R.E., and C.D. Cadena, *Lifespan is unrelated to investment in reproduction in populations of mammals and birds in captivity.* Ecology Letters, 2007. 10(10): p. 867–872.

132 *Selecting birth and death records in Britain and America* Beeton, M., G.U. Yule, and K. Pearson, *On the correlation between duration of life and the number of offspring.* The Royal Society of London Proceedings B, 1900. 65: p. 290–305.

132 *A large-scale contemporary study* Grundy, E., and O. Kravdal, *Reproductive history and mortality in late middle age among Norwegian men and women.* American Journal of Epidemiology, 2008. 167(3): p. 271–279.

132 *A 2006 study* McArdle, P.F., et al., *Does having children extend life span? A genealogical study of parity and longevity in the Amish.* Journals of Gerontology. Series A: Biological Sciences and Medical Sciences, 2006. 61(2): p. 190–195.

132 *Something about bearing a child late in life* Perls, T.T., L. Alpert, and R.C. Fretts, *Middle-aged mothers live longer.* Nature, 1997. 389(6647): p. 133.

132 *Perls did not interpret* Mitteldorf, J., *Demographic evidence for adaptive theories of aging.* Biochemistry, 2009. 77 (7): p. 726–728.

133 *Out of three thousand women* Westendorp, R.G. and T.B. Kirkwood, *Human longevity at the cost of reproductive success.* Nature, 1998. 396(6713): p. 743–746.

133 *I removed just those five* Mitteldorf, J., *Female fertility and longevity.* Age (Dordr), 2010: p. 79–84.

134 *the animals that escaped starvation* McCay, C., M. Crowell, and L. Maynard, *The effect of retarded growth upon the length of life span and upon the ultimate body size.* Nutrition, 1935. 5(3): p. 155.

134 *underfeeding the bionauts had dramatic health benefits* Walford, R.L., et al., *Calorie restriction in biosphere 2: alterations in physiologic, hematologic, hormonal, and biochemical parameters in humans restricted for a 2-year period.* Journals of Gerontology. Series A: Biological Sciences and Medical Sciences, 2002. 57(6): p. B211–24.

134 *A project at Washington University* Holloszy, J.O., and L. Fontana, *Caloric restriction in humans.* Experimental Gerontology, 2007. 42(8): p. 709–712.

135 *CR monkeys were clearly healthier* Mattison, J.A., et al., *Impact of caloric restriction on health and survival in rhesus monkeys from the NIA study.* Nature, 2012.

136 *The monkeys were bored and anxious* Colman, R.J., et al., *Caloric restriction reduces age-related and all-cause mortality in rhesus monkeys.* Nature Communications, 2014. 5.

136 *article claiming to reconcile the CR data with* Shanley, D.P., and T.B. Kirkwood, *Calorie restriction and aging: a life-history analysis.* Evolution: International Journal of Organic Evolution, 2000. 54(3): p. 740–750.

137 *Kirkwood's comparison missed the mark* Mitteldorf, J., *Can experiments on caloric restriction be reconciled with the disposable soma theory for the evolution of senescence?* Evolution: International Journal of Organic Evolution, 2001. 55(9): p. 1902–1905; discussion 1906.

137 *he makes the broadest case possible* Austad, S., *Why We Age.* 1999, New York: Wiley.

140 *cancer rates show no threshold* Luckey, T.D., *Nurture with ionizing radiation: a provocative hypothesis.* Nutrition and Cancer, 1999. 34(1): p. 1–11.

140 *Nevertheless, other longevity factors* Jolly, D., and J. Meyer, *A brief review of radiation hormesis.* Australasian Physical & Engineering Sciences in Medicine, 2009. 32(4): p. 180–187.

141 *Rats have been forced* Calabrese, E.J., *Toxicological awakenings: the rebirth of hormesis as a central pillar of toxicology.* Toxicology and Applied Pharmacology, 2005. 204(1): p. 1–8.

5. When Aging Was Young: Replicative Senescence

146 *life began simple and has grown complex* Lane, N., *Life Ascending: The Ten Great Inventions of Evolution.* 2010: Profile Books.

146 *This has led some scientists to speculate* McFadden, J., and J. Al-Khalili, *Life on the Edge: The Coming of Age of Quantum Biology.* 2015, New York: Crown.

146 *The view that life arrived on Earth* Hoyle, F., and C. Wickramasinghe, *Lifecloud: The Origin of Life in the Universe.* 1978, London: Dent, p. 1.

147 *the extraterrestrial origin* Crick, F.H., and L.E. Orgel, *Directed panspermia.* Icarus, 1973. 19(3): p. 341–346.

147 *The metabolic view starts* Agladze, K., V. Krinsky, and A. Pertsov, *Chaos in the non-stirred Belousov–Zhabotinsky reaction is induced by interaction of waves and stationary dissipative structures.* 1984.

149 *Also with the cell wall came individuation* Dyson, F.J., *Origins of Life.* 1985, Cambridge: Cambridge University Press.

152 *the principle of "multilevel selection"* Wilson, D.S., *Introduction: multilevel selection theory comes of age.* The American Naturalist, 1997. 150(s1): p. S1–S21.

152 *reason that sex became a prerequisite* Mitteldorf, J., *Aging Is a Group-Selected Adaptation.* 2016, Boca Raton, Florida: Taylor & Francis.

153 *Ants do not dominate the earth's biomass* Wilson, E.O., *The Social Conquest of Earth.* 2012, New York: W.W. Norton & Company.

156 *Evolution has prevented the paramecium* Clark, W.R., *Sex and the Origins of Death.* 1998, Oxford: Oxford University Press, p. 208.

156 *the evolutionary meaning of aging* Clark, W.R., *A Means to an End: The Biological Basis of Aging and Death.* 1999, New York; Oxford: Oxford University Press, xv, p. 234.

159 *Hayflick repeated his experiment using cells* Hayflick, L., and P.S. Moorhead, *The serial cultivation of human diploid cell strains.* Experimental Cell Research, 1961. 25(3): p. 585–621.

159 *the ultimate effects of the aging process* Strehler, B.L., *Time, Cells and Aging.* 1977, New York: Academic Press, p. 41.

161 *When telomeres are starting to get short* de Lange, T., V. Lundblad, and E.H. Blackburn, *Telomeres.* Cold Spring Harbor Monograph Series. 2005.

164 *In the study that Cawthon published in 2003* Cawthon, R.M., et al., *Association between telomere length in blood and mortality in people aged 60 years or older.* Lancet, 2003. 361(9355): p. 393–395.

165 *A very large Danish study* Rode, L., B.G. Nordestgaard, and S.E. Bojesen, *Peripheral blood leukocyte telomere length and mortality among 64, 637 individuals from the general population.* Journal of the National Cancer Institute, 2015. 107(6).

6. When Aging Was Even Younger: Apoptosis

169 *The end result of putting him through his paces* Fabrizio, P., et al., *Superoxide is a mediator of an altruistic aging program in Saccharomyces cerevisiae.* The Journal of Cell Biology, 2004. 166(7): p. 1055–1067.

177 *We need apoptosis* Bosco, L., et al., *Apoptosis in human unfertilized oocytes after intracytoplasmic sperm injection.* Fertility and Sterility, 2005. 84(5): p. 1417–1423.

177 *Diseases of old age in which apoptosis is implicated* Marzetti, E., and C. Leeuwenburgh, *Skeletal muscle apoptosis, sarcopenia and frailty at old age.* Experimental Gerontology, 2006. 41(12): p. 1234–1238.

177 *a primary cause of Alzheimer's disease?* Behl, C., *Apoptosis and Alzheimer's disease.* Journal of Neural Transmission, 2000. 107(11): p. 1325–1344.

177 *I described the situation* Mitteldorf, J., *Telomere biology: cancer firewall or aging clock?* Biochemistry, 2013. 78(9): p. 1054–1060.

179 muscle loss: Pistilli, E.E., J.R. Jackson, and S.E. Alway, *Death receptor-associated pro-apoptotic signaling in aged skeletal muscle.* Apoptosis, 2006. 11(12): p. 2115–2126.

179 Menopause: Morita, Y., and J.L. Tilly, *Oocyte apoptosis: like sand through an hourglass.* Developmental Biology, 1999. 213(1): p. 1–17.

180 Alzheimer's disease; Su, J.H., et al., *Immunohistochemical evidence for apoptosis in Alzheimer's disease.* Neuroreport, 1994. 5(18): p. 2529–2533.

7. The Balance of Nature: Demographic Homeostasis

181 *In a famous and influential scholarly article* Slobodkin, L.B., *How to be a predator.* American Zoologist, 1968. 8(1): p. 43–51.

186 *utterly without justification* Williams, G., *Adaptation and Natural Selection.* 1966, Princeton: Princeton University Press.

186 *It was clear to Wynne-Edwards* Wynne-Edwards, V., *Animal Dispersion in Relation to Social Behaviour.* 1962, Edinburgh: Oliver & Boyd.

187 *Huge brown grasshoppers* Wilder, L.I., *On the banks of plum creek.* Little House. Vol. 4. 1937, New York: Harper & Bros.

188 *Forests of the American Midwest* Yoon, C.K., *Looking back at the days of the locust. New York Times.* 2002.

190 *As a consequence* Luckinbill, L., *Coexistence in laboratory populations of Paramecium aurelia and its predator Didinium nasutum.* Ecology, 1973. 54(66): p. 1320–1327.

193 *Such is the relentless logic of exponential growth* Klein, D.R., *The introduction, increase, and crash of reindeer on St Matthew Island.* Journal of Wildlife Management, 1968. 32(2): p. 350–367.

194 *There are powerful and fundamental reasons* Begon, M., C.R. Townsend, and J.L. Harper, *Ecology: From Individuals to Ecosystems.* 2005, New York: Wiley-Blackwell, p. 752.

195 *Even Erwin Schrödinger* Schrödinger, E., *What Is Life?: With Mind and Matter and Autobiographical Sketches.* 1944, Cambridge: Cambridge University Press.

196 *Gilpin had just received* Gilpin, M.E., *Group Selection in Predator-Prey Communities.* 1975, Princeton: Princeton University Press.

201 *In the early computer models* Mitteldorf, J., *Chaotic population dynamics and the evolution of aging: proposing a demographic theory of senescence.* Evolutionary Ecology Research, 2006. 8: p. 561–574; and Mitteldorf, J., and J. Pepper, *Senescence as an adaptation to limit the spread of disease.* Journal of Theoretical Biology, 2009. 260(2): p. 186–195; and Mitteldorf, J., and C. Goodnight, *Post-reproductive life span and demographic stability.* Oikos Journal, 2012. 121(9): p. 1370–1378.

8. So We All Don't Die at Once: Wiles of the Black Queen

203 *a statement of the Demographic Theory of Aging* Mitteldorf, J., *Chaotic population dynamics and the evolution of aging: proposing a demographic theory of senescence.* Evolutionary Ecology Research, 2006. 8: p. 561–574.

206 *In the McGill University laboratory* Hekimi, S., J. Lapointe, and Y. Wen, *Taking a "good" look at free radicals in the aging process.* Trends in Cell Biology, 2011. 21(10): p. 569–576.

211 *In 2006, I demonstrated* Mitteldorf, J., *Chaotic population dynamics and the evolution of aging: proposing a demographic theory of senescence.* Evolutionary Ecology Research, 2006. 8: p. 561–574.

212 *Another sixty years on* Stead, D.G., *The Rabbit in Australia.* 1935, Sydney: Winn.

213 *In 1968, the ecologist Garrett Hardin* Hardin, G., *The tragedy of the commons.* Science, 1968. 162: p. 1243–1248.

216 *I like Carl Woese's answer* Woese, C.R., *Interpreting the universal phylogenetic tree.* Proceedings of the National Academy of Sciences, 2000. 97(15): p. 8392–8396.

217 *a message that could not be ignored* Wagner, G.P., and L. Altenberg, *Complex adaptations and the evolution of evolvability.* Evolution, 1996. 50(3): p. 967–976.

9. Live Longer Right Now

225 *Substances that increase life span when fed to rodents* Knoll, J., *The striatal dopamine dependency of life span in male rats. Longevity study with (–) deprenyl.* Mechanisms of Ageing and Development, 1988. 46(1): p. 237–262; see also: Kitani, K., et al., *Dose-dependency of life span prolongation of F344/DuCrj rats injected with (-)deprenyl.* Biogerontology, 2005. 6(5): p. 297–302.

225 *Rapamycin is the most recent* Harrison, D.E., et al., *Rapamycin fed late in life extends lifespan in genetically heterogeneous mice.* Nature, 2009. 460(7253): p. 392–395.

225 *People who are happy* Diener, E., and M.Y. Chan, *Happy people live longer: subjective well-being contributes to the health and longevity.* Applied Psychology: Health and Well-Being, 2011. 3(1): p. 1–43.

229 *Robert Atkins* Atkins, R.C., *Dr. Atkins' Diet Revolution.* 1972, New York: Bantam.

229 *Barry Sears* Sears, B., *The Zone: Revolutionary Life Plan to Put Your Body in Total Balance for Permanent Weight Loss.* 1995, New York: HarperCollins.

229 *Herman Taller* Taller, H., *Calories Don't Count.* 1961, New York: Simon & Schuster, p. 192.

229 *Two recent studies* Bannister, C.A., et al., *Can people with type 2 diabetes live longer than those without? A comparison of mortality in people initiated with metformin or sulphonylurea monotherapy and matched, non-diabetic controls.* Diabetes, Obesity and Metabolism, 2014. 16(11): p. 1165–1173.

230 *Magnesium is a chemical element* Guerrero-Romero, F., et al., *Oral magnesium supplementation improves insulin sensitivity in non-diabetic subjects with insulin resistance. A double-blind placebo-controlled randomized trial.* Diabetes & Metabolism, 2004. 30(3): p. 253–258; see also: Rodriguez-Moran, M., and F. Guerrero-Romero, *Oral magnesium supplementation improves insulin sensitivity and metabolic control in type 2 diabetic subjects a randomized double-blind controlled trial.* Diabetes Care, 2003. 26(4): p. 1147–1152.

231 *The 120-Year Diet* Walford, R.L., *The 120-Year Diet.* 1988, New York: Pocket Books.

232 *just the* smell *of food* Pletcher, S.D., *The modulation of lifespan by perceptual systems.* Annals of the New York Academy of Sciences, 2009. 1170(1): p. 693–697.

234 *benefits of fasting for cancer patients* Lee, C., et al., *Fasting cycles retard growth of tumors and sensitize a range of cancer cell types to chemotherapy.* Science Translational Medicine, 2012. 4(124): p. 124ra27-124ra27.

234 *primed by starvation for the kill* Longo, V.D., and M.P. Mattson, *Fasting: molecular mechanisms and clinical applications.* Cell Metabolism, 2014. 19(2): p. 181–192.

240 *NSAIDs taken daily are* Kaiser, J., *Will an aspirin a day keep cancer away?* Science, 2012. 337(6101): p. 1471–1473.

10. The Near Future of Aging

249 *Loyola biochemist Phong Le's* Zook, E.C., et al., *Overexpression of Foxn1 attenuates age-associated thymic involution and prevents the expansion of peripheral CD4 memory T cells.* Blood, 2011. 118(22): p. 5723–5731.

249 *There are serendipitous discoveries* Baker, D.J., et al., *Clearance of p16Ink4a-positive senescent cells delays ageing-associated disorders.* Nature, 2011. 479(7372): p. 232–236.

255 *With epithalamin* Anisimov, V.N., and V.K. Khavinson, *Peptide bioregulation of aging: results and prospects.* Biogerontology, 2010. 11(2): p. 139–149.

256 *Arthritis is well-characterized* D'Andrea, M.R., *Add Alzheimer's disease to the list of autoimmune diseases.* Medical Hypotheses, 2005. 64(3): p. 458–463.

256 *Human growth hormone* Fahy, G.M., *Apparent induction of partial thymic regeneration in a normal human subject: a case report.* Journal of Anti-Aging Medicine, 2003. 6(3): p. 219–227.

257 *in 2012 came a dramatic report* Baati, T., et al., *The prolongation of the lifespan of rats by repeated oral administration of [60] fullerene.* Biomaterials, 2012. 33(19): p. 4936–4946.

258 *a signaling pathway that controls just the destructive inflammation* Davis, C.T., et al., *ARF6 inhibition stabilizes the vasculature and enhances survival during endotoxic shock.* The Journal of Immunology, 2014. 192(12): p. 6045–6052.

258 *a target signal called ARF6* Zhu, W., et al., *Interleukin receptor activates a MYD88-ARNO-ARF6 cascade to disrupt vascular stability.* Nature, 2012. 492.7428 (2012): p. 252–255.

265 *a Danish group repeated* Rode, L., B.G. Nordestgaard, and S.E. Bojesen, *Peripheral blood leukocyte telomere length and mortality among 64,637 individuals from the general population.* Journal of the National Cancer Institute, 2015. 107(6).

265 *when Michael Fossel first promoted it* Fossel, M., *Reversing Human Aging.* 1997, New York: HarperCollins.

265 The Immortal Cell *by Michael D. West* West, M.D., *The Immortal Cell.* 2003, New York: Doubleday, p. 244.

265 *The University of Texas laboratory* Baur, J.A., et al., *Telomere Position Effect in Human Cells.* Science, 2001. 292(5524): p. 2075–2077.

268 *In the Spanish laboratory of* Bernardes de Jesus, B., et al., *The telomerase activator TA-65 elongates short telomeres and increases health span of adult/old mice without increasing cancer incidence.* Aging Cell, 2011. 10(4): p. 604–621.

269 *These experiments were resurrected* McCay, C.M., et al., *Parabiosis between old and young rats.* Gerontology, 1957. 1(1): p. 7–17.

269 *In the first study from the Conboys* Conboy, I.M., et al., *Rejuvenation of aged progenitor cells by exposure to a young systemic environment.* Nature, 2005. 433(7027): p. 760–764.

270 *Blood plasma from young mice* Katcher, H., *Studies that shed new light on aging.* Biochemistry (Moscow), 2013.

270 *This is not a practical* Villeda, S.A., et al., *Young blood reverses age-related impairments in cognitive function and synaptic plasticity in mice.* Nature Medicine, 2014.

270 *An energy enzyme* Katsimpardi, L., et al., *Vascular and neurogenic rejuvenation of the aging mouse brain by young systemic factors.* Science, 2014. 344(6184): p. 630–634.

272 *proposed that epigenetics constitutes an independent aging clock* Johnson, A.A., et al., *The role of DNA methylation in aging, rejuvenation, and age-related disease.* Rejuvenation Research, 2012. 15(5): p. 483–494; and Rando, T.A., and H.Y. Chang, *Aging, rejuvenation, and epigenetic reprogramming: resetting the aging clock.* Cell, 2012. 148(1): p. 46–57; and Mitteldorf, J., *How does the body know how old it is? Introducing the epigenetic clock hypothesis.* Biochemistry (Moscow), 2013. 78(9): p. 1048–1053.

274 *For example, setting back* Goldman, D.P., et al., *Substantial health and economic returns from delayed aging may warrant a new focus for medical research.* Health Affairs, 2013. 32(10): p. 1698–1705.

11. All Tomorrow's Parties

281 *But the sixth extincton* Kolbert, E., *The Sixth Extinction: An Unnatural History*. 2014, New York: Henry Holt and Company.

282 *E. O. Wilson's well-reasoned, alarming prediction* Wilson, E.O., *The Meaning of Human Existence*. 2014, New York: Liveright Publishing.

284 *Donna Haraway of UC–Santa Barbara* Haraway, D., *Staying with the Trouble: Making Kin in the Chthulucene*. 2016, Durham: Duke University Press, forthcoming.

289 *century of accelerating change* Kurzweil, R., *Human life: the next generation*. New Scientist, 2005. 24: p. 32–37.

293 *Stephen Cave* Cave, S., *Immortality: The Quest to Live Forever and How It Drives Civilization*. 2012, New York: Crown.

INDEX